FIRE AND EMERGENCY SERVICES ADMINISTRATION

MANAGEMENT AND LEADERSHIP PRACTICES

L. CHARLES SMEBY, JR., M.P.P., MIFIREE

JONES AND BARTLETT PUBLISHERS

Sudbury, Massachusetts

BOSTON   TORONTO   LONDON   SINGAPORE

Jones and Bartlett Publishers
World Headquarters
40 Tall Pine Drive
Sudbury, MA 01776
978-443-5000
info@jbpub.com
www.jbpub.com

Jones and Bartlett Publishers Canada
6339 Ormindale Way
Mississauga, Ontario L5V 1J2
Canada

Jones and Bartlett Publishers
International
Barb House, Barb Mews
London W6 7PA
United Kingdom

Jones and Bartlett's books and products are available through most bookstores and online booksellers. To contact Jones and Bartlett Publishers directly, call 800-832-0034, fax 978-443-8000, or visit our website www.jbpub.com.

Substantial discounts on bulk quantities of Jones and Bartlett's publications are available to corporations, professional associations, and other qualified organizations. For details and specific discount information, contact the special sales department at Jones and Bartlett via the above contact information or send an email to specialsales@jbpub.com.

Production Credits
Chief Executive Officer: Clayton E. Jones
Chief Operating Officer: Donald W. Jones, Jr.
President, Higher Education and Professional Publishing: Robert W. Holland, Jr.
V.P., Sales and Marketing: William J. Kane
V.P., Production and Design: Anne Spencer
V.P., Manufacturing and Inventory Control: Therese Connell
Publisher, Public Safety Group: Kimberly Brophy
Acquisitions Editor: Bill Larkin
Associate Editor: Janet Morris
Production Editor: Jenny L. McIsaac
Director of Marketing: Alisha Weisman
Interior Design: Anne Spencer
Cover Design: Timothy Dziewit
Composition: Auburn Associates, Inc.
Cover Photograph: © Jones and Bartlett Publishers. Photographed by Kimberly Potvin.
Text Printing and Binding: Malloy Inc.
Cover Printing: Malloy Inc.

Library of Congress Cataloging-in-Publication Data

Smeby, L. Charles.
  Fire and emergency service administration : management and leadership practices / L. Charles Smeby.— 1st ed.
    p. cm.
  ISBN 0-7637-3189-7
  1. Fire departments—Management. 2. Leadership. I. Title.
  TH9158.S64 2006
  363.37068'4—dc22
                                                    2005024504

Printed in the United States of America
09 08 07 06 05   10 9 8 7 6 5 4 3 2 1

Contents

Chapter 10 Ethics   146

Chapter 11 Public Policy Analysis   155

Foreword

Mission Statement

"To mitigate the threat to life and property from, fire, medical and other emergencies through education, prevention, community preparedness, emergency response and recovery programs." –Colorado Springs (Colorado) Fire Department.

Fire and emergency service organizations are entering a new era of expanded and professional emergency services. Many organizations were suddenly propelled into a new level of consciousness after the September 11, 2001 terrorist attack.

In addition to the many new responsibilities for fire departments such as EMS, hazardous materials, fire prevention, and disaster planning, traditional fires continue to occur. Although fewer fires occur now than in the past, these incidents remain a challenge, perhaps even more so since fire fighters have less real world experience with fighting fires. This has led to a disturbing trend. According to the National Fire Protection Association (NFPA) special report on fire fighter deaths, "While the number of structure fires has dropped, the rate of fire fighter deaths at structure fires has not."

The fire and emergency services profession has made many progressive changes in equipment, training and funding, but clearly the job is not finished. The job of making improvements is never finished. This text provides the knowledge for chief officers, and their staff, to identify and implement progressive change that will help keep fire fighters safe while providing the best public fire and emergency services.

Some Advice

Throughout this text, the words of other writers are used to support the ideas and observations of the author. The future is built upon the past. The following is some good advice to help the reader with the proper mindset for reading and understanding this text:

- Take care of yourself. Good health is everyone's major source of wealth. Without it, happiness is almost impossible.
- Resolve to be cheerful and helpful. People will repay you in kind.
- Avoid angry, abrasive persons. They are generally vengeful.
- Avoid zealots. They are generally humorless.
- Resolve to listen [and read] more and to talk less. No one ever learns anything by talking.
- Be cautious of giving advice. Wise men don't need it and fools won't heed it.
- Resolve to be tender with the young, compassionate with the aged, sympathetic with the striving and tolerant of the weak and the wrong. Sometime in life you will have been all of these.
- Do not equate money with success. There are many successful moneymakers who are miserable failures as human beings. What counts most about success is how a person achieves it. [Especially true in the fire and emergency services] (Lloyd Shearer, 1989)

Biography

MR. SMEBY is currently a lecturer at the University of Florida. He develops, creates, and presents courses and provides administrative support for the Fire and Emergency Service Bachelor's distance learning program. He was an academic instructor at Florida State Fire College for six years.

Previously, he was a Senior Fire Service Specialist at the National Fire Protection Association in the Public Fire Protection Division. He coordinated several NFPA fire service technical committees and provided guidance to the implementation, understanding, and the intent of NFPA fire service standards.

Before joining the NFPA, Mr. Smeby had a twenty-year career with the Prince George's County, Maryland, Fire Department, where he retired as a Battalion Chief.

The Past

Knowledge Objectives

- Comprehend the history of fire and emergency services and its impact on contemporary organizations.
- Understand the effect that the insurance industry has had on both building and fire prevention codes.
- Recognize the influence the insurance industry (Insurance Service Office) has had on the location and placement of fire stations, equipment, and staffing of present day fire departments.
- Examine the influence of the National Fire Protection Association (NFPA) on the fire service.
- Comprehend federal involvement in regulation and funding of state and local fire and emergency services.
- Examine progressive trends in fire and emergency services.

Prologue to the Future

There is an old saying that reminds us that if we do not study and understand history, we are bound to make the same mistakes as those that have occurred in the past. It is also true that we should identify what works so we can expand and continue proven, high-quality practices.

Often progress comes from trying out new, unproven ideas. We applaud all the courageous and imaginative administrators who tried and failed, but kept trying, learning from their failures, and finally succeeding. These are the unsung heroes of our profession—the visionaries, risk takers, and true leaders.

Fire Service History

Administration, leading change, and management all become more meaningful when you know the history as well as the current status of the fire and emergency services profession. It is like going on a journey—the map must show where to start (the past) as well as the final destination (the future or goal), and all the points in between.

The first known organized fire protection in North America began in 1648 in New Amsterdam, a Dutch colonial town on the island of Manhattan that was later renamed New York City by the British. These early fire departments were all-volunteer organizations.

Massive fires would strike many large cities in the United States, and they were the stimulus for many changes to the U.S. fire service. Demands for stricter fire safety codes and paid fire departments were common after each conflagration (**FIGURE 1-1**).

The responsibility for fire protection in the United States has continued to be local, in large part due to our federal form of government. The national government regulates the fire services to some extent in those 25 states that have state Occupational Safety and Health Administration (OSHA) enforcement plans. In these

FIGURE 1-1 Early American Firemen

states, any new OSHA safety regulation becomes enforceable for state and municipal employees.

In the United States today, there are estimated to be 30,542 fire departments. Because there are relatively few federal regulations for fire services in the United States, this presents the possibility of 30,542 different ways to provide public fire protection. Volunteer fire fighters staff a majority of these fire departments, slightly less than half (14,817) serving small populations of 2,500 or less. Presently, all-paid departments protect 44.8% of the U.S. population, all-volunteer departments protect 21.5%, and 33.7% are protected by combination departments (NFPA, 2004). What is surprising is the relative uniformity in equipment and procedures among the fire services in the United States.

However, differences still exist that keep many departments from helping each other during major emergencies. These differences can be either organizational or equipment associated. For example, in 1991 a major wildland-urban interface fire struck Oakland, California; by the time it was finally extinguished, it was the largest property loss fire in U.S. history. Twenty-five people died, and 2,950 homes along with 450 apartments burned during this fire. Oakland requested mutual aid from surrounding jurisdictions; however, when the mutual aid fire engines arrived and tried to connect their hoses to the Oakland fire hydrants, they found that the Oakland connections were a different size, hampering their ability to help.

Another very common difference is that many departments use different radio channels and therefore are unable to communicate with each other. We still have room for improvement and change. There is always something that can be done better tomorrow.

America Burning

The *America Burning* report to the President of the United States, along with its recommendations, has been the basis for many of the changes in progressive fire and emergency services organizations (USFA, 1973). One example of progress was the result of a recommendation on page 167, Appendix IV. The report states: "The Commission recommends that fire departments, lacking emergency ambulance, paramedical and rescue services consider providing them, especially if they are located in communities where these services are not adequately provided by other agencies." Virtually all fire departments providing service to populations of 100,000 or more people provide emergency medical services (EMS) and this trend continues upward (NFPA, 2005).

Many of the newer National Fire Protection Association (NFPA) standards are a direct response to recommendations contained in this report. Starting in the early 1980s, NFPA standards for "professional qualifications" were created for fire service members. In addition, several standards have been created to cover hazardous materials response personnel and, more recently, rescue personnel. And of course, there is NFPA 1500, *Standard on Fire Department Occupational Safety and Health Program*, which covers training and safety issues.

America Burning is still seen as very significant and timely—so much so that the U.S. Fire Administration (USFA) has reprinted it along with a short "Information Update." Although some of the statistics have either changed or are new, including more accurate information that has recently become available, many of the observations and recommendations in this report are still relevant to present-day fire and emergency services organizations.

The history of any profession is important to understand because it explains many of the current policies and may point out issues that have never been resolved. It is the job of an administrator to help lead the organization to be the best it can be within legal and/or financial restraints. However, both legal and financial restraints can be changed using information from the present and past to help explain the problems and justify potential solutions.

Insurance Service Office (ISO) Grading Schedule

Most of the uniformity of practices among American fire departments that exists today is a direct result of the Insurance Service Office (ISO) *Grading Schedule for Municipal Fire Protection*. This document was created by the National Board of Fire Underwriters, an organization whose members come from the insurance industry. The catalyst for this fire department evaluation standard was several major fires and conflagrations that occurred in the United States from the late

1800s to the early 1900s. The insurance industry needed a method to encourage fire departments to be prepared for these large fires and to have the ability to give and receive mutual aid for departments that could not handle a major fire on their own.

Several insurance companies went bankrupt because they could not pay all of their claims after these large fires. Other insurance companies formed associations that attempted to pay the claims not paid by the bankrupted insurance companies. There was financial panic in the insurance industry as a result of these citywide fires.

A conflagration in Baltimore on February 7, 1904, caused the mayor of the city to request help from several major cities including Washington, D.C., Philadelphia, and New York. When out-of-city fire companies arrived in Baltimore, they laid out their hoses and discovered, to their dismay, that their hose couplings did not fit the Baltimore fire hydrants. Without water from hydrants, the mutual aid fire companies were severely handicapped and the city continued to burn. The insurance fire claims were excessive and several insurance companies were unable to pay for all their losses.

After this fire, the National Board of Fire Underwriters created the grading schedule and started using it to survey the public fire protection departments in urban centers throughout the United States. After over 90 years of surveying both large and small cities, towns, and counties, a great amount of uniformity now exists. Most U.S. fire departments now use a uniform coupling and threads or have adapters for their fire hoses that enable compatibility.

Municipalities were encouraged to adopt the recommendations from the surveys or face increases in fire insurance costs. Municipalities are graded by ISO with a numerical rating from Class 1 to 10, with 1 being the best rating; a community that has a Class 1 rating will have lower insurance premiums than a community with a higher rating. Many fire chiefs (administrators) have justified budgetary increases for items recommended by ISO to improve a community's rating, thus lowering insurance costs.

Recently, some fire officials and municipal administrators have been openly critical of the grading schedule, which was originally created to prevent large-scale conflagrations. To receive a high rating, a fire department would have to invest heavily in many fire companies. As the probability of large-scale fires decreased, public officials frequently questioned the need for large crews (up to six fire fighters per company) and close spacing for fire stations, sometimes as close as 3/4 mile apart.

National and regional building and fire codes, adopted locally or statewide, also had their beginnings after these large fires. These building and fire codes also began with the insurance industry, and have had a substantial impact on reducing loss of life and property in building fires, practically eliminating the conflagration potential from U.S. cities.

Early Fire Prevention Codes

A leader in the area of national consensus codes, the NFPA was founded in 1896. Insurance companies were also the organizers of this association. The first standard created addressed uniform automatic fire sprinkler installation.

Around 1913, NFPA added a standard for *Safety to Life,* which addresses a continuing concern among the American public to prevent fire deaths. This was a direct reaction to several multi-fatal fire tragedies in the United States.

NFPA creates, updates, and adopts standards using a consensus method of committees that represent the fire protection community. Membership is open to any interested person from insurance companies; federal, state, or local governments; fire departments; manufacturers of fire equipment; the public; and other fire protection interests. Each committee has members that represent a variety of separate interest groups.

Today's U.S. Fire Service

The following are paraphrased excerpts from the conclusion of a recent study of the U.S. fire service concerning its needs and response capabilities (FEMA/ NFPA, 2002):

- There is a great need to refurbish or replace many fire stations, as well as to purchase new fire apparatus and additional self-contained breathing apparatus (SCBA).
- Many fire companies arrive with fewer than four personnel, indicating a need to recruit or hire more members.
- Many departments are not able to provide all of the fire prevention services that would help reduce fire deaths and property loss.
- Most of the revenues for all- or mostly volunteer fire departments come from taxes.
- A majority of fire fighters serve in fire departments with no program to maintain basic firefighter fitness and health.
- In communities with a population of fewer than 2,500 and a fire department staffed

mostly by volunteers, there are only four or fewer fire fighters available to attend to a mid-day house fire.

- The majority of fire departments do not have all their personnel certified to basic levels of EMS training.
- An estimated 233,000 fire fighters, mostly volunteer, lack formal training in fire-fighter duties.
- An estimated 42% of the U.S. population is protected by fire departments that do not have a program for free distribution of home smoke alarms.

New Trends in Fire Service Standardization

Only 14 fire departments in the United States protect more than 1 million people, and the largest department is actually a consolidation of five counties (New York). On the other end of the spectrum, 14,817 departments protect fewer than 2,500 people each (Karter, 2005). The U.S. fire service could be characterized as being made up of many small organizations with a few that protect larger cities or counties. Therefore, the U.S. fire service, which is now over 350 years old, is still trying to develop substantial agreement on issues such as, "What is quality professional fire service?" The term *professional service* in this text means those organizations and members who are trained and certified to professional qualification standards as promulgated by recognized national or international fire and emergency service organizations.

In the past 25 years, NFPA standards have been created for fire-fighter training and education, equipment, and emergency operations. This is a national trend to standardize firefighting operations and safety in the United States.

A survey on April 30, 2001, asked the following question: "Should the fire service fall under NFPA standards or OSHA standards?" **TABLE 1-1** lists the results.

If you total the results of the NFPA only and NFPA and OSHA responses, you would find that 86.5% of the respondents would favor NFPA standards to regulate and guide the fire service. On top of that, most OSHA standards are based on or are NFPA standards.

Table 1-1 *Firehouse* Magazine Internet Survey	
NFPA only	44.1%
OSHA only	6.3%
Both NFPA & OSHA	42.4%
Neither	7.1%
(Firehouse.com, 2001)	

This leaves only 7.1% of the survey participants who are not in favor of these standards. This was the result of 7,431 votes cast at the Firehouse.com website (2001). Although not a statistically random study, the fire ser-vice is clearly looking to the NFPA for guidance. Remember, it is not NFPA the organization that creates these standards, but committees made up of fire ser-vice subject matter experts with the input of the public through an open consensus process.

Planning Tools

Many fire departments have relied on the ISO grading schedule for justifying and planning improvements to their departments. On March 15, 2001, the lead article for *IAFC OnScene* proclaimed, "State Farm abandons ISO grading schedule, fire chiefs wonder, 'What now?'" Chief Mike Brown, President of the International Association of Fire Chiefs (IAFC), asked the following question, "What will the fire chiefs [administrators] use as a tool for budgetary planning to improve their fire departments and maintain them at the highest possible level?"

Other insurance companies are sure to follow. State Farm is changing to a system that determines the insurance rate based solely on the loss experience in what they call *Subzone Rating Factors*. These factors include fire, wind, hail, water damage, theft, and liability. For example, if it pays out $20 million during a year, it wants to charge rates that recover the payouts along with a percentage profit. State Farm has stated that 70% of claims paid under its homeowner's program are nonfire losses, although it has not produced any hard data to support this claim.

As an alternative to ISO for planning, the IAFC and the International City/County Management Association (ICMA) have created a new *accreditation* program for fire/rescue agencies. It is a comprehensive self-analysis of each agency's ability to provide professional fire and emergency services to the public. This self-analysis is submitted to a committee that evaluates the competencies using professional standards as guidelines. This committee is appointed by the IAFC and ICMA and includes individuals who have extensive knowledge and experience in public fire protection. This is a positive step in the right direction; however, it is possible to be accredited and have different levels of staffing and response times as compared to other accredited departments.

For EMS, the Commission on Accreditation of Ambulance Service administers an accreditation program that was created in 1991. Again, this is a comprehensive self-evaluation analysis with the main objective being to provide quality and timely emergency medical services to the community.

Recently, an NFPA committee finished work on a new standard, NFPA 1710, *Standard for the Organization and Deployment of Fire Suppression Operations, Emergency Medical Operations, and Special Operations to the Public by Career Fire Departments*, which measures fire and EMS deployment, response times, and staffing. Chief Brown, president of the IAFC, had the following comments; "I recommend using NFPA 1710 as the future benchmark for fire and emergency services. I also recommend using the Commission on Fire Accreditation International (CFAI) self assessment process to measure the capabilities of our fire departments, to measure our response times capabilities to fire and emergency medical service calls and to measure our compliance with NFPA 1710 and other NFPA nationally recognized standards" (IAFC, 2001).

Finally, there are many consultants who can be hired to study and make recommendations on any aspect of a fire and emergency services organization. This will be discussed in more detail in Chapter 11.

New Trends

The *NFPA Fire Protection Handbook*, 19th edition (2003, p. 7-7), points out "In recent years, the role of the fire service in many communities has expanded far beyond fire suppression." The name *fire department* doesn't begin to cover the many services that progressive organizations are providing to their communities. For example, the NFPA reported that 61% of the calls to fire departments in 2003 were medical aid calls. Public safety is the business of the modern fire department. With this expansion, fire prevention and public education have appropriately begun receiving an increased emphasis as the proactive elements of a fire service delivery system. Citizens are dependent on the fire department to ensure their protection against the dangers of fire, entrapment, explosion, and any emergency event that may occur in the community. "Recent acts of domestic terrorism in Oklahoma City, Washington D.C., and New York City have added a new mission to the fire service" (NFPA, 2003, p. 7-7). The job of the administrator is becoming more complicated and challenging every day.

More and more services are being requested of a traditional fire service organization at the same time that many revenue sources are becoming stagnant or declining. The *NFPA Fire Protection Handbook* states,

"Quality management stresses a total organizational approach that makes the quality of service, as perceived by the customer, the number one driving force for the operation of the Organization" (NFPA, 2003, p. 7-10). To use this technique of evaluation and guidance, the fire service must first answer the question, "Are our customers capable of judging our services?"

Customer Satisfaction

In the business community, customer satisfaction is of great importance to those businesses that are dedicated to quality service and gaining a bigger share of their market. To determine customer satisfaction levels, the customer has to be able to judge the service or product. In the fire and emergency services business, however, most customers are not really able to judge the quality of the services provided. In spite of this lack of understanding of emergency services, a 1997 survey reported in *USA Today* stated that the fire service was "trusted" by 78% of the citizens surveyed. (In contrast, the police were trusted by 46% of respondents, public schools 32%, local TV news 24%, city or local government 14%, state government 9%, and federal government 6%.) The public in general seems to trust the fire service and believes it is doing a good job. However, this loyal trust does not always equate to quality service or adequate funding.

This approval also can be a double-edged sword. Because the public generally thinks the fire services are doing a good job, it may be difficult to convince them that additional resources are needed. However, if a convincing argument is presented, backed up by strong facts and data, the public generally will trust that the requests are justified.

It is easy to assume that excellent service is being provided and changes are not needed because there is generally very little quality feedback from customers. In most cases, the administrator will be fully responsible for measuring the fire department's competency. This evaluation process will require a great amount of courage, especially for controversial but progressive changes that may be needed. Great administrators are not afraid of risk and have the courage to go forward, even when faced with substantial opposition.

Federal Involvement

At the present time, federal agencies such as the Federal Emergency Management Agency (FEMA) along with the Department of Defense and the Federal Bureau of Investigation (FBI) are providing training, equipment, and resources to prepare local fire and emergency services organizations for terrorist activities, weapons of mass destruction, and other extremist violence. This federal effort will continue to expand as a result of the September 11, 2001 terrorist attacks on the United States.

In addition, the Environmental Protection Agency (EPA), a major regulator of hazardous substances, is enforcing its regulations with increasing efficiency. It has been very effective in creating new requirements for safer storage, use, and transportation of hazardous materials.

The original driving force behind the increasing interest in hazardous materials was the growing concern for worker safety and health. OSHA has developed rules that help focus attention on the needs of employees and on the obligations of their employers. Labor unions and litigation have helped bring about appropriate adherence to the many and varied OSHA regulations. Concern for worker safety has also resulted in the realization that emergency responders need, and are entitled to, adequate protection against the hazardous materials that they come into contact with in their jobs.

At the federal level, a national focus on fire protection is struggling to stay alive. The USFA and its National Fire Academy (NFA) have done an excellent job providing a forum to enhance the ability of the fire service to deal with fire problems. The NFA provides training and education for the U.S. fire service, but its impact has been limited as a result of insufficient federal budget support. However, as a result of the September 11th terrorist attacks, the fire service is beginning to get increased national attention, and the influence and prestige of the National Fire Academy is growing.

These federal programs provide a path to national uniformity by stressing curriculum, ideas, management, and leadership practices that have proven to be successful.

U.S. Forest Service

Established in 1905, the Forest Service is an agency of the U.S. Department of Agriculture. The Forest Service manages public lands in national forests and grasslands. This federal agency has developed two very innovative programs.

The first is the National Interagency Incident Management System (NIIMS), also called the incident command system. All wildland fire protection agencies are organized to handle a reasonable number of forest,

brush, and grass fires within their jurisdictions. They usually can fight most fires with their own resources plus aid from other nearby agencies. However, additional or substantial outside assistance may be required for larger destructive wildfires, especially when they threaten homes built in the wildland-urban interface. NIIMS provides a total systems approach for response in such emergency situations. NIIMS was developed in the early 1970s and was essential for the management of large-scale wildland fire incidents in southern California.

Today, NIIMS is a system for responding to a wide range of emergencies, including fires, floods, earthquakes, hurricanes, tornadoes, tidal waves, riots, spilling of hazardous materials, and other natural or human-caused incidents. NIIMS includes five major subsystems (command, operations, planning, logistics, and finance), which together provide a comprehensive approach to incident management.

The second program is colloquially called *red card*. This is a printout of the current wildland fire qualifications of an individual. Fire fighters assigned to a wildland fire being managed by a federal agency (U.S. Forest Service, Bureau of Land Management, National Park Service, Bureau of Indian Affairs, or U.S. Fish and Wildlife) or by many state agencies are required to have a red card. In a sense, it is similar to a driver's license. The credentials specify levels of competency, including firefighting and incident command, achieved by successfully completing training and testing to a national standard.

Deputy Chief Smith stated, "This overhead team [from the National Interagency Fire Center] came in from Texas, Arizona, Montana and New Mexico, among other areas." He further commented, "In my 30 years of involvement with structural firefighting [Washington, D.C.], I had never before witnessed a complete staff of educated, experienced and credentialed commanders mitigating an incident" (Smith, 2003). The red card system of credentialing works quite well. For example, a wildland fire fighter from North Carolina can travel to Wyoming and be assigned to a team made up of individuals from all over the country because there is a national system of competency credentialing.

Emergency Services Incident Management System

An organization called FIRESCOPE (FIrefighting RESources of California Organized for Potential Emergencies), involving all agencies with firefighting responsibilities in California, was organized after the disastrous 1970 wildfires in southern California. The goal of this group was to create and implement a system to manage and coordinate these large-scale emergencies. More recently, FIRESCOPE, the Phoenix (AZ) Fire Department, the U.S. Forest Service, the USFA, and the NFPA formed a discussion group to create a standard incident management system (IMS) for all types of emergencies.

The 2002 edition of NFPA 1561 contains the latest consensus by this consortium of fire service organizations. The changes help resolve issues of terminology, modular expansion, implementation at the emergency scene, and multi-agency, multi-jurisdictional command conflicts. The national implementation of the IMS is still a work in progress, but has now attracted the support of the federal government as the best way to manage and mitigate terrorist and other natural and manmade disasters. This is critically important when many different emergency services departments must work together at these tragic national events.

As outlined in the NFPA 1561 standard, incident scene management is critical to the successful mitigation of both large- and small-scale emergencies and the safety of the emergency responders. IMS is critical for fire fighter occupational safety and health. "The absence of incident command at an incident scene puts fire fighters at great risk and is one of five leading contributing factors of fire fighter fatalities, as reported by NIOSH" (NFPA, 2003, p. 7-75).

On March 1, 2004, the Department of Homeland Security (DHS) released its National Incident Management System (NIMS). Although many of the details have not been completed, this incident management system will probably become the default system in the United States. At this time, there are several national incident management systems used in this country. The varying characteristics and terminology make it very difficult for two agencies using different systems to operate efficiently together at the scene of a major emergency. Full and complete implementation will take many years, but the emergency services now have a single incident command system.

Tom Ridge, Secretary of DHS, explains, "This unique system (NIMS) provides all of our Nation's first-responders and authorities with the same foundation for incident management, in terrorist attacks, natural disasters, and other emergencies. From our Nation to our neighborhoods, America is safer" (DHS, 2004).

Current U.S. Fire Service Trends

Smoke alarms are working; they have saved many lives. Sadly, fire fatalities continue to occur in some homes with smoke detectors because of missing or dead batteries. The following statistics are from the NFPA's research conducted in 2004:

- Fifteen of every 16 homes (96%) in the United States have at least one smoke alarm.
- Roughly half of home fire deaths result from fires in the small percentage of homes with no smoke alarms.
- Homes with smoke alarms (whether or not they are operational) typically have a death rate that is 40%–50% less than the rate for homes without alarms.
- In one-quarter of the reported fires in homes equipped with smoke alarms, the devices did not work. Households with nonworking smoke alarms now outnumber those with no smoke alarms.
- Why do smoke alarms fail? The most common reason behind smoke alarm failure is a missing, dead, or disconnected battery. (NFPA, 2004)

More trends:

- "Automatic sprinklers, long used for property protection, are now entering the life safety arena and have been installed in hotel rooms, apartments, and homes. Although somewhat new, success rates are extremely high. *America at Risk, America Burning Recommissioned* (2000) found that 'The most effective fire loss prevention and reduction measure with respect to both life and property is the installation and maintenance of fire sprinklers.' A report conducted over 15 years in Scottsdale, Arizona, reported no fire deaths in homes protected with fire sprinkler systems (41,408). In the same time period, 13 people died in unprotected homes. In addition, fire damage was substantially less, at $2,166 per incident versus $45,019 in unsprinklered incidents. In Scottsdale, the cost of sprinklers was only $.80 per square foot in 2001" (Home Fire Sprinkler Coalition, 2003).
- Fire safety education, which started in the 1970s, has now become part of many of our nation's schools and fire departments. A new trend is the *all risk* approach for safety education by NFPA and FEMA. In addition to fire

safety, many other safety concerns are included in these programs such as vehicle accidents, pool and water safety, and other common accidents that cause injuries and deaths.

- Greater standardization among fire departments is being driven by NFPA standards such as 1500, *Standard on Fire Department Occupational Safety and Health Program,* and 1710, *Standard for the Organization and Deployment of Fire Suppression Operations, Emergency Medical Operations, and Special Operations to the Public by Career Fire Departments.*
- More mutual aid agreements are being signed to prepare for major emergency incidents that could result from terrorist actions and catastrophic natural disasters. At the National Fire and Emergency Services Dinner on April 30, 2003, DHS Secretary Ridge stated, "And we're developing a national incident management system that emphasizes mutual aid. And this is something we need to stress over and over again, a plan that emphasizes mutual aids across jurisdictional and geographic lines, centered on an incident command system you [the fire service] develop and use across the nation. And again, we need to take a look at some of our colleagues around the country, take a look at the mutual aid systems that they have in California, in Florida, and other states that actually have statewide mutual aid agreements. We need to work together to make sure all 50 states have state-wide mutual aid agreements."
- National credentialing for structural fire fighters and incident commanders is being developed under the encouragement of DHS. At a meeting of the National Fire Academy (NFA) Board of Visitors on April 28, 2003, Dr. Denis Onieal, the Superintendent of the NFA, discussed the need to develop a system of credentialing: "The response to 9-11 is an example that displayed this need. All that was needed at these response sites was a fire coat, and they were allowed access to the disaster site. It was recognized that a national Incident Management System would be useful—a national system of identifying people who respond. A committee has been formed to design this system. One of the concepts that will be included is if a firefighter is going to be considered for a Federal response, he/she would have to have some form of identifica-

tion (ID) that allows him/her to be there. The ID would include credentials."

- Careless smoking habits are a leading cause of fire deaths, but these deaths are decreasing thanks to the increased effectiveness of smoke alarms and a decrease in the number of smokers. Recently, Philip Morris (a major producer of cigarettes) proposed creating a self-extinguishing cigarette to address this critical fire safety problem.
- Fire departments are placing more emphasis on fire prevention, especially in modern building and fire codes.
- Fire departments are adding new services such as rescue, hazardous materials, emergency preparedness, and emergency medical services.

Emergency Preparedness/ Management History

FEMA, the U.S. government agency tasked with responding to, planning for, recovering from, and mitigating disasters, began in 1803 with a congressional act that was designed to assist a New Hampshire town following a catastrophic fire. In the next 200-plus years, the federal government created many other programs to provide funding for recovery from disasters. These programs were placed in many different federal agencies; for example, flood control was placed under the U.S. Army Corps of Engineers.

During the 1960s and 1970s, major disasters such as hurricanes and earthquakes brought new focus onto the existing fragmented approach to federal disaster assistance (**FIGURE 1-2**). In 1979, FEMA was expanded to consolidate the federal response to disasters, which at the time included the U.S. Fire Administration and the Civil Preparedness Agency in the Defense Department.

With the end of the Cold War, FEMA could redirect resources from civil defense to disaster relief, recovery, and mitigation programs. This agency has started programs to prevent and reduce the risk before a disaster strikes. FEMA is now focusing on preparing and delivering training to mitigate and recover from terrorist incidents.

TABLE 1-2 lists disasters for which emergency responders must be prepared to provide prevention, mitigation, and recovery from the tragic consequences.

In most cases, the local fire and emergency services organization is the first to respond to these dis-

FIGURE 1-2 Hurricane Floyd, 1999

asters. Therefore, when large-scale disaster strikes, the federal emergency preparedness effort must complement and assist the local efforts.

U.S. Department of Homeland Security

The U.S. Department of Homeland Security was established on November 25, 2002, to protect our country and its citizens from terrorist attacks. This new department includes FEMA as one of its key components. FEMA will continue efforts to reduce life and property loss from all types of hazards and to help communities and citizens avoid becoming victims of disasters.

FEMA will be the nation's incident manager for all hazards and major disasters. This will provide a direct line of authority and communications from the president, via the Secretary of Homeland Security,

down to the local level. "But we face in many ways many of the same challenges, trying to break down traditional ways of looking at things in the post-9/11 environment, and getting people on board to help set emergency management priorities. The issues we spoke about a year ago, mutual aid, information sharing, emergency credentialing, interoperability, security clearances, secure video teleconferencing are just as important today" (Ridge, 2003).

The number one priority of the new Department of Homeland Security is to prevent an attack from happening. This is also very important to fire and EMS providers, because it has always been more effective to prevent problems than to respond to an emergency in progress. Traditionally, however, prevention has not been a major priority with most emergency services.

Ridge commented, "Your police, your fire departments, your emergency medical technicians—we know you're the first ones [called]. We know that if something happens in your community, they don't hit the phone and dial area code 202—they're not dialing Washington, D.C. But they dial 911, they dial the folks at home." He continued, "On September 11th, 2001, 347 of your colleagues died with the collapse of the World Trade Center, but 99 percent—99 percent of the people below the crash lines got out alive. Ninety-nine percent of all those people got out alive thanks to fire fighters and emergency responders" (Ridge, 2003).

In responding to future disasters, a plan that details mutual aid across jurisdictions, states, and nationally, using the incident management command system will be a top priority. In addition, "It is imperative that we provide ways for you to communicate [primarily two way radio] across jurisdiction lines [a common failing in every major disaster]" (Ridge, 2003).

Table 1-2 Natural and Manmade Disasters	
• Building fires	• Landslides and mudflows
• Chem-bio attack	• Nuclear power plant emergency
• Disruption of telephone and radio communications	• Terrorism
• Earthquakes	• Thunderstorms and lightning
• Extreme heat	• Tornadoes
• Floods and flash floods	• Tsunamis
• Hazardous materials	• Volcanoes
• Hurricanes	• Wildland fires
• Internet computer attacks	• Winter storms

EMS History

Organized EMS existed in early military medicine through the use of horse-drawn carts to remove casualties from the battlefield during the Napoleonic wars. A civilian ambulance service was first seen in the United States in the 1860s. These ambulances were hospital-based and transported patients to the hospital for medical treatment. In the past, it was also common in many communities for the local funeral home to use its hearse to transport patients to the hospital.

The modern EMS started with the 1966 publication of *Accidental Death and Disability: The Neglected Disease of Modern Society* by the National Academy of Sciences' National Research Council. The report served as a blueprint for a national effort to improve emergency medical care. Found in the report were the following items:

- Ambulance services showed a diversity of standards, which were often low; frequent use of unnecessarily expensive and usually ill-designed equipment; and generally inadequate supplies.
- There was no generally accepted standard for the competence or training of ambulance attendants.
- For decades, the emergency facilities of most hospitals consisted only of accident rooms—poorly equipped, inadequately staffed, and ordinarily used for limited numbers of seriously ill persons or for charity victims of disease or injury.

One of the greatest federal impacts on EMS was the creation of the National Highway Traffic Safety Administration in the Department of Transportation.

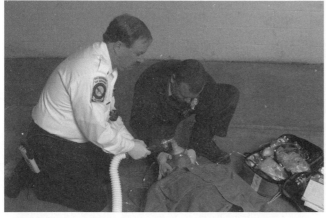

FIGURE 1-3 Paramedics in Action

This agency was responsible for creating training curricula for prehospital care providers such as first responders, EMTs, and paramedics (**FIGURE 1-3**). These training curricula are the de facto standard for competency of EMS providers in this country.

Through this federal agency, funding became available for training, equipment, and planning. At about the same time, a number of military medics were coming back from the Vietnam War. They had experience with professional emergency medical training, organization, and equipment that had not been seen in any other military operations. This emergency medical military system was also very successful at saving the lives of critically wounded soldiers. For example, the use of helicopters for fast transport to emergency medical treatment was standard operating procedure in Vietnam.

Emergency—The TV Show

In 1971, the television show *Emergency* debuted and contributed to changing the general public's attitude concerning the fire service and emergency medical care. At the start of the show, there were only 12 medic units in the entire country. Four years later, at least 50% of the population of this country was within 10 minutes of a medic unit. This television program had more to do with the fast adoption of advanced and basic emergency medical service by the fire service, hospitals, and ambulance services than anything else. The reruns of this show are still being broadcast and have excellent audience ratings even today, over 30 years later:

> The [TV] show "EMERGENCY!" has done a lot for the careers and desires of the paramedics in the field. The show generated interests within the community to spawn a mobile-based care unit, as well as peaking the desire of many young people to become paramedics themselves. Who would have guessed that the humble beginnings of the Mobile Intensive Care Unit would evolve to one of today's most recognized emergency based services, and one that is used most routinely by millions of people. (Project 51, 1999)

An Example of Progress in the Fire and Emergency Services

In 1994, a state fire marshal made inquiries to find out if there was a fire safety standard that covered casino ships. He was in the process of reviewing plans for a new casino ship that would dock in his state and sail on the Mississippi River. The NFPA did not have a standard at that time. (Subsequently, at the request of the U.S. Coast Guard, the NFPA did create a fire safety standard.) While reviewing the plans, he noticed that the permit applicant was the previous owner of the Beverly Hills Supper Club in Southgate, Kentucky. This nightclub was the location of a tragic fire in 1977 that took the lives of 165 innocent people.

He asked himself how this person, who was at least partially responsible for many fire deaths, could be allowed to own and operate a ship that needs to be safe enough to protect the lives of thousands.

As he began to think about this situation, he remembered another tragic fire that took place more recently, in 1991. In Hamlet, North Carolina, a fire broke out at a meat processing plant and claimed the lives of 25 workers. The owner was convicted of 25 counts of manslaughter, lost all his wealth in civil court action, and spent many years imprisoned. This is in sharp contrast to the consequences encountered by the owner of the Beverly Hills Supper Club some 14 years prior.

There has been a great deal of progress in fire and emergency services, but because it is relatively slow and not always obvious to the casual observer, the following sarcastic comment about the fire service has been heard: "200 years of tradition unhampered by progress!" However, this overstatement is definitely not true. **TABLE 1-3** lists some examples of relatively recent progressive changes in the fire and emergency services.

Generation Challenges

Some characteristics of younger generations are challenging for many fire and emergency services administrators; solutions will be addressed in the following chapters. "Our future fire fighter will have precious little experience with teamwork, self-sacrifice, personal organization and respect for authority. It's not necessarily his or her fault. Young people have grown up in the soft, affluent society that was created for them. We sold our farms to corporations and won the Cold War so our kids didn't have to bale hay or get drafted" (Page, 1999). What kind of leadership will be needed in the future based on our present and past experiences with our members?

"The real question is whether the fire service should change to accommodate those characteristics or whether we should expect individuals to adapt to traditional fire service standards [teamwork, self-sacrifice, personal organization and respect for authority]?" (Page, 1999). This text will try to answer this question.

Table 1-3 Progressive Changes in Fire and Emergency Services	
• Smoke detectors	• Safety standards for fire fighters
• Hurricane- and earthquake-resistant construction	• Defibrillators
• Paramedic services	• Federal support for disaster mitigation
• Professional fire fighter and officers competency standards	• NFPA standard for the deployment and staffing of fire companies
• Federal grants for fire and emergency services	• Thermal imagers
• Compressed air foam	• Hydraulic rescue tools
• National standards for emergency medical training and certification	• National focus of emergency response on major disaster events
• Higher education opportunities	• Chem-bio detectors
• National Fire Academy	• FEMA and the USFA
• Incident management system	• Seat belts and air bags
• Residential fire sprinklers	

CASE STUDY #1: An Example of Progress

In 1981, an engine company was dispatched to a report of an explosion at a major university's chemistry building. They were first due and the company officer mentally reviewed his knowledge of hazardous materials. He had not completed any formal classes in the handling of hazardous materials incidents, and NFPA 472, *Standard for Professional Competence of Responders to Hazardous Materials Incidents*, would not be created for many years. (This standard contains the competencies that are commonly used to create training courses and certify hazardous materials response personnel today. The requirements contained in this standard have been incorporated into OSHA and EPA regulations that are mandatory for all fire and emergency services personnel that respond to hazardous materials incidents.)

The officer had studied several textbooks on hazardous materials for the lieutenant's promotional exam and was an avid reader of all fire service periodicals. During officer candidate school, several lectures were presented by chief fire officers who had completed a community college hazardous materials course.

As they approached the university chemistry building, the driver hooked up to the fire department standpipe system and the officer and crew entered the building. All crew members were equipped with full protective clothing and SCBAs. They ascended the stairs to the third floor and, before opening the door on the floor of the reported explosion, stopped and noticed a vision panel (wire glass) in the fire door.

The officer thought that this was very fortunate because he could observe the floor without actually entering the corridor and its potentially hazardous atmosphere. It was at that time that he noticed a university campus police officer crossing the corridor from a classroom laboratory to the bathroom where the victim was located. With this observation, he was now convinced that it was safe to enter the corridor as he had just seen the police officer working without any obvious symptoms of distress. They entered to find the victim of the blast without placing their SCBA facepieces on.

About 5 years later, this same fire department had an official hazardous materials response team and formal training. It was a locally created program, again before NFPA 472, but it was very comprehensive and recognized nationally as being very progressive. It was at this time that they became aware that the adverse effects of hazardous materials might not show up until years after an exposure. Therefore, the safety of the atmosphere in that chemistry building could not have been reliably determined except by scientific testing equipment.

One summer afternoon in 1986, a hazardous materials rescue squad from this department was dispatched to a possible chlorine leak. This was not an ordinary chlorine installation, but was the central bulk storage for a major water theme park.

When the rescue squad arrived, they found an engine and a ladder company were already on the scene. The fire fighters had evacuated all the water rides, had cordoned off hot and warm zones, and were waiting for the arrival of the hazardous materials technicians in a safe location, upwind of the storage facility. They had not attempted an entry. This was real progress. When the hazardous materials team went in to survey the interior of the water treatment building, they had full protective clothing and were using SCBAs.

Close to the end of the officer's career in the fire department, two of his close friends had retired and within a year's time were diagnosed with inoperable cancerous tumors in their brains. They were both in their late 40s when they died. These two cases help to point out the possibility that the adverse consequences of exposures to toxic products might not appear until many years later. The operations in the chemistry laboratory back in 1981 could have exposed the crew to a toxic substance that may not have had any medical symptoms until many years after the exposure.

Hazardous materials responders today are very fortunate to have credible operations and safety standards that can be used as the basis for their training, education, safety, and certification. These standards are the achievement of a select group of subject matter experts that are the members of an NFPA Hazardous Materials Response Personnel Committee along with the review and input from the fire and emergency services community. They represent many different interests and views that together have formed a synergistic consensus process that has produced these standards. This is progress.

Discussion Questions

1. In a case study format, document a department's past history and its present situation in being able to extricate victims from vehicle crashes. In many cases, this will require personal interviews with several senior members. You may even have to locate some retired members.

2. Is there a formal training program for members who operate extrication equipment today? If yes, please describe.

3. Comment on any changes that have occurred in this department since NFPA 1006, *Standard for Rescue Technician Professional Qualifications*, was created.

CASE STUDY #2: A Second Example of Progress—Maybe?

As explained previously, the U.S. Department of Homeland Security has promulgated a National Incident Management System. Even before it was announced, there were those in the fire and emergency services community who opposed this specific system. Some noted that it was not designed to address a multi-alarm fire command system that was typically used in certain parts of the country.

Many of the details of this system have not been completed at the time this book went to press. Common terminology for resources must be created. In the federal system, if you need 100 widgets, everyone must know what a widget is and be able to provide them if they are part of the system. This is not a trivial task.

One simple example is the definition of the basic unit of a fire department: the engine company. The name of this unit can vary by department, such as engine, pumper, wagon, or unit. One state plan has the following definition of a Type I Engine: 1,000 GPM pump, 750 gallons of water, 1,200 feet of supply hose, 200 feet of hand lines, and four fire fighters. Each fire fighter on a Type I Engine must be state certified as a Fire Fighter II.

The qualifications and certifications for personnel are also left as future additions to the plan. Currently, there is no consensus or consistency in the fire service for a system to determine competency or credentialing of fire personnel. Although a few states have mandatory fire fighter certification systems, the majority of fire departments operate independently, and each decides how to train its members.

Before answering the following discussion questions, please download and review the following: http://www.dhs.gov/interweb/assetlibrary/NIMS-90-web.pdf and http://www.usfa.fema.gov/downloads/pdf/publications/fa-282.pdf.

Discussion Questions

1. Do all fire and emergency services organizations have to comply with the federal emergency management plan?

2. Would your department's engine companies measure up to the Type I Engine requirements listed above? If not, would your department change, or do you think that the requirements are not realistic and should be changed? Please provide a complete justification for your answer.

3. After reading the DHS document, are there any major differences between that plan and your department's standard operating procedures for incident management? If so, please list the differences.

4. After reading the FEMA document, is your fire and emergency services agency ready for a large catastrophic event? Please provide a detailed point-by-point analysis.

References

Department of Homeland Security (DHS). *Department of Homeland Security Director Tom Ridge Approves National Incident Management System (NIMS)*. http://www.dhs.gov/dhspublic/display?content=3259 (accessed July 21, 2005), 2004.

FEMA. *America at Risk, America Burning Recommissioned.* Emmitsburg, MD: FEMA, 2000.

FEMA. *Responding to Incidents of National Consequence: Recommendations for America's Fire and Emergency Services Based on the Events of September 11, 2001, and other similar incidents.* http://www.usfa.fema.gov/about/media/2004releases/052604.shtm (accessed July 21, 2005), 2004.

FEMA/NFPA. *A Needs Assessment of the U.S. Fire Service* 2002. http://www.nfpa.org (accessed July 21, 2005), 2002.

Firehouse.com. "Should the fire service fall under NFPA standards or OSHA standards?" http://www.firehouse.com/polls/2001/042301_body.html (accessed November 5, 2001), 2001.

Home Fire Sprinkler Coalition. *Fire Sprinkler Facts.* http://www.homefiresprinkler.org/hfsc.html (accessed May 8, 2003), 2003.

IAFC. "Testimony of Chief Brown." *IAFC OnScene.* International Association of Fire Chiefs, July 12, 2001, pp. 1–3.

Karter, M.J., Jr. *U.S. Fire Department Profile through 2003.* Quincy, MA: NFPA, 2005.

NFPA. *1710, Standard for the Organization and Deployment of Fire Suppression Operations, Emergency Medical Operations, and Special Operations to the Public by Career Fire Departments.* Quincy, MA: NFPA, 2001.

NFPA. *472, Standard for Professional Competence of Responders to Hazardous Materials Incidents.* Quincy, MA: NFPA, 2002a.

NFPA. *1500, Standard on Fire Department Occupational Safety and Health Program.* Quincy, MA: NFPA, 2002b.

NFPA. *NFPA Fire Protection Handbook*, 19th ed. Quincy, MA: NFPA, 2003.

NFPA. *Smoke Alarms.* http://www.nfpa.org/categoryList.asp?categoryID=278&URL=Research%20&%20Reports/Fact%20sheets/Fire%20protection%20equipment/Smoke%20alarms (accessed July 21, 2005), 2004.

Page, James O. "Your 21st century firefighter." *Fire-Rescue Magazine.* November 1999, p. 10.

Project 51. *History of the TV show* Emergency. http://www.squad51.org (accessed June 23, 2001), 1999.

Ridge, Tom. Remarks to the Congressional Fire Service Caucus (April 30, 2003) and National Emergency Managers Association (September 15, 2003), 2003.

Smith, Michael L. "Wildland & WMD: More in common than you think." *Firehouse,* September 2003, pp. 29–30.

USA Today. Whom do you trust? http://archives.usatoday.com (accessed July 21, 2005), 1997.

USFA. *America Burning—The Report of the National Commission of Fire Prevention and Control.* http://www.usfa.fema.gov/usfapubs/pubs_main.cfm, 1973.

2

Introduction to Administration

Knowledge Objectives

- Define administration and its function.
- Understand how management and leadership are an integral part of administration.
- Recognize the influence politics, the public, and special interest groups have on fire and emergency services administration.
- Understand how to gain influence over the political process.
- Discuss how to build a team for effective administration.
- Examine how rules and regulations, commonly called standard operating procedures (SOPs), affect the consistency and effectiveness of emergency services.
- Understand the differences, advantages, and similarities of career and volunteer service agencies.
- Discuss how an effective administrator gains honest and expert advice from both outside sources and peers.

What Is Administration?

Effective administration requires two skills: management and leadership. Although there is some overlap between the two, the following example should help clarify the differences.

At the scene of a structural fire, to be successful at extinguishing the fire, fire fighters need the following general items: adequate staffing, equipment, standard operating procedures (SOPs), and training. Management will provide the people, equipment and training given a budget plan and direction. This is done prior to the emergency and is part of the planning, organizing, and procurement function of the organization; in other words, it is management that prepares the organization to be able to accomplish its goals. This makes management essential to the organization's ability to achieve its goals.

Leadership at the company level can be observed when trained personnel safely and efficiently complete their mission. Using the resources provided by the fire and emergency services organization, the company officer is responsible for leading a company of fire fighters in successfully and safely conducting emergency operations. Leadership in this role is a combination of training, experience, and courage.

There is another level of leadership higher up at the organization level, and this is the type of leadership that will be discussed in this text. In most cases it takes leadership skills to be able to "sell" the need for new programs, additional resources, and/or progressive changes. Once the resources become available or the new change is implemented, it then becomes the job of management to use them efficiently and effectively to support the mission.

Management and Leadership

A person who excels in management skills will be very good at making the existing organization work properly and efficiently. Generally, management-oriented people represent the majority of administrators in fire and emergency services organizations. American companies have been very good at training great managers who have propelled our country to the top of the industrialized world. The Walt Disney Company, for example, was created and led by a great visionary, Walt Disney. The present administration of Disney focuses more on keeping profit levels high for the stockholders. Although this sequence of events does not occur in every organization, and both past and current Disney leaders have showed both leadership and management, this sequence has been common in many industries. In many cases, the people who select chiefs are looking for someone who will be a good caretaker of public funds, staff, and resources.

However, managers may not be proficient at leadership skills such as vision and risk taking. Physical risk taking is a trait that appears to be inherent in some individuals such as fire fighters, soldiers, and athletes. But in administration, it is a skill that can be learned. A person with strong leadership skills will have a vision of where the organization needs to go and the courage to attempt the journey. There may be no guarantee that the leader will arrive at the visionary goal, but because leaders are risk takers, they try anyway.

It is possible to gain an understanding of and competency in both types of skills, even if the administrator feels more comfortable in one of the roles than the other. The person who is leadership-oriented may have difficulty remembering the necessity of having people with strong management skills running the day-to-day operations. A leadership-oriented administrator will focus on a visionary goal 5 to 10 years down the road, but may forget to plan for tomorrow. In contrast, a management-oriented person generally will not be comfortable with risk taking and would rather not get involved in visionary goals that have no guaranteed outcome. Although these descriptions may be oversimplifications, many administrators feel more comfortable using one style more frequently than the other. However, they can learn some leadership or management skills through formal education, self-study, attending seminars, and experience.

One of the techniques a manager uses to reduce the anxiety of risk taking for a truly innovative change is to start a program but make it voluntary. This empowers members or low-level supervisors to do whatever they would like to do to reach the new goal (or not reach the goal). This will foster inconsistencies, however, because supervisors will set their own timetables for implementation.

A leadership-oriented person is more of a risk taker when it comes to new ideas and change. Many people drawn to the fire and emergency services have a reduced sense of fear and are risk takers in physical activities. There is an old saying that highlights the heroic behavior of fire fighters: When fire fighters are entering a burning building; everyone else is leaving it. However, when it comes to changes in the organization, fear of the unknown may make many fire and emergency services administrators hesitant.

Organizational Monopoly

Public fire and emergency services are essentially monopolies either provided by a government or allowed to operate without control by the local government (e.g., independent volunteer departments). This is a unique way to look at public emergency services, and many people will have to think about this idea for a while before it becomes comfortable for them. In the big picture, most municipal departments have no trouble understanding the issue of government oversight, but many independent volunteer companies would strongly resist any oversight by elected representatives. A monopoly is a situation where only one provider or company has complete control of the market. This means that the business can charge prices that are above what would be expected in an open market with many competing businesses. In the United States, monopolies are illegal unless they are a government-regulated industry or service.

If a house catches fire, for example, the occupants do not have to go to the yellow pages to choose a fire department to call. Fire and emergency services will be provided to anybody who calls, in any area, but only one primary provider will be responsible for the service provided.

This is a strong justification used by elected officials who represent the customer to regulate emergency services. This same justification is used to provide and regulate the local law enforcement agency.

In addition, it is rare for the citizens to be able to satisfactorily judge the quality of the emergency service, so it is expected and assumed that elected representatives and their staffs will have this knowledge. This is also true for other public services such as public roads, police, and schools, which are also provided solely by government, and are hence monopolies. In addition, many of the utilities, such as electricity, gas, telephone, and cable television, are monopolies and are regulated by local, state, or federal governments.

The Impact of Politics on the Fire and Emergency Services Organization

When you hear the term *politics*, does it bring up a negative feeling of uncaring elected officials taking care of special interest groups? In your community, do the big money contributors have more influence than the ordinary citizens? Whatever the case, although some individual elected officials may abuse the power entrusted to them, this does not mean that the entire political system is not legitimate. Remember that politics is the foundation of the United States' representative form of government.

Although political systems are not perfect, many elected officials are heavily influenced by the local voters. Most elected officials have many goals, but one goal that is fairly common is *to get re-elected*. In most cases, this means taking care of the voters' and/or the campaign contributors' needs.

Many voters are concerned about the amount of taxes they pay. You can therefore count on elected officials questioning any request for additional funding for the department. These politicians are acting as the voice of the individual taxpayer who may believe that taxes are already too high. This is especially common in jurisdictions with large numbers of taxpayers who are on fixed incomes such as Social Security. This group of voters will continue to increase in numbers with the retiring "baby boomer" generation.

Sources of Political Power
There are two sources of political power: formal authority and informal external groups of influential citizens. The administrator should understand and be in touch with members of the latter groups, as well as any influential administrators or staff in other departments. Their influence cannot be underestimated. They can make or break attempts at change or to gain increases in the budget.

The following are sources of formal authority:
- Legislative branch: Elected officials
- Executive branch: Appointed officials
- Judicial branch: The judges
- Other federal, state, or local agencies that may have oversight or regulatory responsibilities

The following are external power sources:
- Employee or member associations (e.g., unions and volunteer fire companies)
- Media
- Special interest groups
- Professional organizations
- Business and industry
- Homeowners' groups
- Social clubs
- Religious organizations

This is not an exhaustive list, but it can provide a good starting point for building support for a change in policy, additional staffing, or new equipment/facilities. In addition, the general public may help with the approval of any proposed changes if they find some position to support.

As mentioned in Chapter 1, a large percentage of citizens trust the fire service and believe it is doing a good job. By continuing to demonstrate actions that show fairness, honesty, responsiveness, openness, and accountability, your agency will build an even stronger support base with the general public, elected and appointed officials, and influential citizens in your jurisdiction.

Characteristics of the Political Process
The following is a short list of some items administrators should assess before attempting a policy change:
- What political party is in control now?
- Are the elected officials short-term or long-term thinkers, or some of each?
- Are there any elected officials expected to lose or not run for re-election in the upcoming elections? (This can have a positive or negative effect.)
- Are elected officials elected at-large or from districts?
- Are there any signs of taxpayer revolt?
- Is there any expectation of what is *politically correct?*
- Are there any elected officials who would have a reason to not support the fire and emergency services organization?
- Are there any officials who would like to be a *champion* or strong supporter of the fire service?

- Do any of these elected representatives *owe* a special interest group that may oppose your changes? (Remember, employee groups, developers, home builders, and the like are entitled to donate funds to re-election campaigns.)
- Has a single-issue candidate recently been elected? (For example, to find as much tax money as possible to improve schools.)

The Political Challenge

Do not allow the unique constraints or challenges of politics to keep the department mired in mediocrity. Be something more than just a caretaker. Any attempt to implement change will take courage, sacrifice, and determination.

The difference between caution and courage can be the difference between the status quo and successful progressive change. Always taking the safe path and insisting on having all the answers before making a decision will keep the administrator from leading change in any organization. Seeking new frontiers for your vision and thriving on the unpredictability of the future will help achieve real change. Courage and vision will reduce the unexpected changes that tend to push many organizations into crisis management mode. It is relatively easy to be a hero at the emergency incident; it takes a lot more determination to be a courageous leader who ventures out to make significant change in the organization.

Remember that there are always some people who will be against any change simply because it is change. If you do not feel comfortable with some criticism, you are in the wrong business. As General Colin Powell has said:

Being responsible sometimes means making some people angry. . . . Good leadership involves responsibility to the welfare of the group [and public], which means that some people will get angry at our actions and decisions. It's inevitable— if you're honorable. Trying to get everyone to like you is a sign of mediocrity: You'll avoid the tough decisions, you'll avoid confronting the people who need to be confronted, and you'll avoid offering differential rewards based on differential performance because some people might get upset. Ironically, by procrastinating on the difficult choices, by trying not to get anyone mad, and by treating everyone equally "nicely" regardless of their contributions, you'll simply ensure that the only people you'll wind up angering are the most creative and productive people in the organization. (Powell, 1996)

The following are some skills needed to overcome political opposition:
- Accept and understand the legitimacy of politics and elected officials.
- Understand the structure and process of politics and government.
- Build political alliances.
- Obtain the support of special interest groups.
- Acquire public support through effective marketing.
- Practice open and honest communications with employees, citizens, media, and appointed and elected officials.
- Use conflict resolution, negotiation, and bargaining techniques.
- Identify the various stakeholders and any benefits they may receive as a result of your change; this can help develop strategies for enlisting the support of these influential individuals and groups.
- Develop trust (from members in your organization, the public, media, elected representatives, and appointed officials).

Administrative Tools for Accountability

The following is from the National Fire Protection Association (NFPA) *Fire Protection Handbook,* 19th edition:

Rules and Regulations
As with any organization, rules and regulations are needed to govern operations. This is especially true in the fire service, due to the hazardous nature of much of the activity and the need for a clear understanding of expected performance.

Every fire department should have a set of rules and regulations that outline performance expectations for its members, the standard operating procedures for the department, and disciplinary action that may be taken for failure to follow the regulations. These rules and regulations can be, and often are, supplemented by orders from the fire chief who may add to or clarify the rules or change them for a special event or specific purpose. Both the rules and regulations

Administration duties in a fire and emergency services organization involve analyzing, reviewing, creating, deciding on, and enforcing rules and regulations for the organization. A few of these rules may be mandated by a higher authority in government or comply with state or federal regulations. These rules may also be the result of a task force of members empowered by the administrator to create SOPs for safety and efficiency during emergency operations.

There has been controversy about the use of SOPs for emergency operations. These are essentially rules and regulations. Some argue that if these SOPs are put in writing, then the organization becomes liable for failures to comply with the SOPs. The answer to this concern is very straightforward: If SOPs are promulgated and then enforced rigorously and consistently, there will not be any liability. Vigorous enforcement will actually help protect the organization from lawsuits.

Rules and Regulations

Enforcing rules and regulations is one of the most difficult duties that an administrator may have, other than requesting additional funding from elected officials. There will always be some members and supervisors in fire and emergency services organizations who do not like rules and regulations. However, for accountability and consistency, rules are absolutely essential, especially during emergency operations. To add to the difficulty, even a great charismatic leader will have a small percentage of people who will intentionally or inadvertently fail to comply with rules and regulations. This job takes diligence and perseverance.

Why are rules and regulations necessary? After a devastating hurricane in Florida, newly constructed buildings were required to use construction techniques to reduce or eliminate the damage from these powerful storms. Building and fire codes are rules and regulations that administrators can use to accomplish their mission. For the administrator, it is easier to find justification and public support for strict enforcement and changes after a major disaster.

As a result of a fire fighter's death from smoke inhalation, an SOP was issued that required the use of self-contained breathing apparatus (SCBA) at all interior structural fires (**FIGURE 2-1**). Again, this was relatively easy to mandate in the organization that had the

FIGURE 2-1 Fire Fighter Using a Self-Contained Breathing Apparatus

death; however, other fire and emergency services organizations may experience resistance from members.

Experiences such as these are the driving force behind new requirements in national consensus standards, which may be adopted by local, state, or federal governments as mandatory regulations. It is always better to learn from the mistakes of others than to have to learn from your own.

These types of rules and regulations, NFPA standards, and Occupational Safety and Health Administration (OSHA) regulations are relatively new to the fire service. It is well accepted that this process of developing national safety standards for the construction industry (i.e., building and fire codes) is valid and universally used.

Although it is clear that fire service members support using these standards and regulations, fire and emergency services organizations and their administrators are still in the process of implementing the changes necessary to comply. Some fire and emergency officials along with elected and appointed municipal officials outwardly resist complying with standards and regulations if they personally disagree or if they have a fiscal impact. An April 2001 article in *Fire Chief* magazine asked four safety officers, "Has your department done a safety audit as described in the appendix to NFPA 1500 [*Standard on Fire Department Occupational Safety and Health Program*]? What were the results?" None of the safety officers interviewed reported that they were in full compliance with this safety standard that was first adopted by the NFPA in 1987, fourteen years prior to the interviews. These safety officers estimated anywhere from 60% to 90% compliance (*Fire Chief*, 2001).

There also can be operational, staffing, training, and communications inconsistencies in the fire and emergency services, which all add up to differences from agency to agency, town to town, and state to state. This adds to the public's difficulty in judging the service provided and determining if it is truly receiving professional-quality emergency service. A sign on the side of an engine that proclaims FIRE DEPARTMENT does not guarantee the same level of service for all similarly marked fire trucks throughout the country. Although the fire service has made a lot of progress, this is still work in progress.

Even in some metro fire and emergency services organizations, the service levels can vary by shift and/or battalion. Without rules and regulations, freelancing and independent goal setting become the norm, which can be deadly on the scene of a major fire. The public deserves an emergency services organization that can operate with a high degree of consistency of behavior and an unusual degree of conformity with prescribed patterns of action. Hence, the fundamental discipline and conformance to SOPs is valuable and should be very similar to military operations.

A consensus process can be used to determined public policy, mission, SOPs, or goals, using the input from task groups of members representing all levels from fire fighter to chief. Once SOPs are adopted, the only way to achieve a high degree of conformity and consistency throughout the organization is with rules and regulations or SOPs that are actively enforced. This is especially true for those decisions that must be made in emergency situations when first arriving responders have only 5 to 15 seconds to consider all the alternatives.

However, there should also be some room for exceptions based on the judgment of the officer. The assumption is that the officers have been through a rigorous process of training, education, and certification. An appropriate system to critique these emergency field decisions will assure that this latitude is not used without justification.

For example, at a fire in a large vacant mansion, the first due engine stretched its hose line to the front door and began to enter. The department's SOPs required the first due engine to attack the fire from a position on Side 1 (the front). This is where the front door is found in most buildings. But on entering they noticed, luckily through the heavy smoke, that the floor had burned through and there was a large hole into the basement that was filled with heavy fire. They quickly retreated and radioed this information to the Battalion Chief. Plan B was then implemented.

Leaders vs. Administrators

In *America Burning* (USFA, 1973, p. 20), a description of the situation in the fire services at that time states, "Presiding over this tenuous alliance [fire department] is the fire chief, who wears two hats—one, the administrative hat required to run the organization; the other, the helmet he dons when the alarm is sounded to lead his fire fighters in the suppression of a fire. Since the fire chief usually has come up through the ranks, the second hat probably fits comfortably. It is the administrative duties of today's complex municipal department for which the chief is less likely to be adequately prepared." This national report goes on to say, "The typical hiring and promotion system in which everyone from the chief on down started as a rookie fireman has guaranteed good leaders who understand the needs of the men under them" (p. 36).

America Burning used the term *leadership* when discussing the ability to lead fire fighters at the scene of an emergency incident. In this text, the term *leader* is used at the organizational level, which calls for a special set of skills and knowledge.

Leaders are problem solvers. That's why this text discusses the many aspects of *problem solving*. More precisely, it's about the very special challenges to problem solving that fire and emergency services administrative officials encounter in the public sector.

Mission-Driven Bureaucracy

The most dedicated, innovative, and creative public administrators may fail under the frustrations of pressure from special interest groups and elected or appointed officials. This is an example of what can be called *good people in a bad system*. In many organizations that are not focusing on their mission, this lack of focus is usually caused by a defective organization and its administrative rules. Some of these rules are not in writing but are commonly called tradition.

For example, a career fire department was staffing its engine companies with two or three members. At fires, it was common for the officer to be the person operating the nozzle. Apparently this was done because of the lack of adequate staffing, and because it was a privilege given this position (tradition). With the arrival of a newly appointed chief from the outside, staffing increases were attained, which brought all engine and truck companies up to four members.

However, the traditional engine company operations were not changed. This turned out to be a contributing factor in the deaths of two fire fighters and their lieutenant. With the lieutenant on the nozzle, and the two fire fighters behind, the officer was not able to supervise the operation and was not aware of the deteriorating situation. Their hoseline burned through, leaving them without any water to control the fire. Because the hoseline had been extended down one flight of stairs into a basement, the fire also burned through their only known avenue of escape—the stairs leading up to the ground-level floor.

Many organizations are driven not by their missions, but by their rules (traditions) and their budgets. They have a rule for everything that could conceivably go wrong and a line-item for every category of spending. Too many meticulous administrative rules can restrict an organization from being mission- or customer-focused.

As a general guideline, rules can be broken down into two major categories: management and emergency operations. As mentioned previously, for emergency operations, basic rules and regulations are absolutely necessary to provide consistent, safe, and efficient fire and emergency services, especially to cover operations in the first few minutes.

In a fire and emergency services organization, there are always some people or special interest groups that want more rules and some that do not want any. Employees who are unionized will want to have a written contract that contains many rules on personnel management. Items such as when overtime is to be paid and many other conditions of employment are typically included in a labor agreement. Most of these agreements are necessary to prevent inconsistent and unfair actions in personnel management. However, some contain items that can restrict effective administration.

Volunteer Fire and Emergency Services

Many volunteer fire and emergency services organizations are independent private corporations that are operated without formal scrutiny by the municipality or public (customers) they serve. Independent fire and emergency services organizations are often completely self-regulated without accountability to anyone but themselves. In these cases, accountability is left to volunteer members who are elected as chiefs and/or a group of elected members named commissioners or board of directors. These people owe their election to those they must supervise. These are well-meaning, dedicated people who really want to do nothing more than serve their communities. However, this situation can lead to a conflict of interest.

It is very difficult for volunteer organizations that elect their own officers to have the same stability and consistency as a municipal fire and emergency services organization. This is not a result of bad or incompetent people; it is a result of a system in which officers are elected by a democratic process. When new volunteer chiefs are elected, they may have a lot of latitude in enforcing, interpreting, and changing the SOPs, training requirements, equipment, and apparatus needs and specifications. Volunteer chiefs could initiate change each time a new chief is elected. In addition, as soon as they are elected they must start planning on re-election, which can temper their ability to be a strong leader. Typically, they are elected for 1-year terms.

A similar situation exists in the U.S. House of Representatives. These representatives are elected on 2-year terms. Almost as soon as they are elected and take office, they must start planning for their re-election. This causes them to spend a considerable amount of time and effort on re-election strategies and takes away from their key mission, which is to represent the people in their district and the interest of the United States. In most cases, these are honorable people who are placed in a system that has limitations. This is one of the reasons that the federal budget keeps getting larger as each representative sees a need to bring federal dollars back to their home district. This allows them to take credit, and improves their chance of re-election. This is not the sole driving force behind all their actions and votes in Congress, but it is always on their minds.

In many volunteer organizations, an annual election of the officers produces less consistency from year to year, from chief to chief, or among different organizations throughout the country. With a lack of rules (or standards), some of these organizations can drift from providing excellent to very poor service. The majority of these organizations give good service and many have the potential to improve their service by using the appropriate set of professional standards.

Volunteer organizations seem to work best when the elected officials for the area served appoint the chief and appropriate rules and regulations are created and enforced. Fire and emergency services organizations are granted a monopoly (either formally or by default) to provide a critical emergency service, and there should be some form of accountability for the service performed by representatives of the public they serve.

Career Fire Chiefs

Just as there are limitations to volunteer organizations that elect their officers, there are problems when elected officials or county/city administrators appoint chiefs. One problem involves the loyalty issue that pulls this person generally among three separate groups—the elected/appointed officials, the employee's labor group, and the public (customer).

The chief has to attempt to maintain a professional independent position. At times this can make the job a very lonely one. What is truly difficult about this situation is that the two groups that many times have a self-interest—elected/appointed officials and the labor organization—will be the most powerful. The other group, the public, will have the weakest direct influence because of their lack of knowledge about emergency services. It is this group that the fire and emergency services chief should be an advocate for in a controversial debate.

It is not uncommon to find fire and emergency services administrators who feel alienated from their organizations. The chief may feel that the labor organization, the elected officials, and other appointed administrators all want to convince or force their particular idea(s) on the organization. In the role of looking out for the best interests of the public, the chief should discuss and/or research any new ideas to determine their validity or reasonableness before adoption and implementation.

A simple example will illustrate this predicament. A proposed incentive pay for employees who have been certified as paramedics is being discussed. This change will probably be supported by the union and opposed by the elected and appointed officials. The public may have no opinion because this issue will likely not receive enough notoriety via the local press for them to become aware of the situation.

A proposed agreement between the municipal administration and the labor organization stipulates that funds for this incentive pay are to be taken from a new program that the chief had championed last year and that had been successfully funded: NFPA's Risk Watch, an educational program for young children intended to reduce and eliminate loss of life and property damage from accidents. "For children ages 14 and under, the number-one health risk isn't drugs or disease: it is injuries. Each year, unintentional injuries

kill more than 6,000 kids and permanently disable more than 120,000" (NFPA, 2004). The chief is now squarely in the middle, and could simply go along with this agreement by keeping quiet, knowing that the public would not be aware that they had lost a proven safety program.

Chapter 3 provides a full discussion of methods that can help chiefs gain the influence needed to best resolve these conflicts with a solution that keeps the public's needs in mind. In addition, Chapters 4, 10, and 11 have helpful discussions related to this problem. These chapters present methods that give fire and emergency services administrators the best chance of having their professional judgment prevail.

Ensuring Professional Administrators

Both career and volunteer organizations may have difficulty selecting the best person for the chief's position. Whether the person is elected or appointed, the result may be the same: The administrator (chief) may be ineffective. For example, in an April 2004 editorial, Janet Wilmoth of *Fire Chief* magazine quoted a fire chief who said ". . . spineless, mealy-mouth fire chiefs are a barrier to enforcing rules and regulations that can and will save fire fighters lives" (Wilmoth, 2004) Fire and emergency services are evolving into a profession, and the selection of chiefs should be similar to other professions such as doctors, engineers, and lawyers.

Briefly, these professions need three areas of competency: training, experience, and education. These competencies should meet minimum levels of proficiency and be the same throughout the country. For example, a medical doctor must complete college-level education as well as medical school, and then an internship/residency (training and experience). Only after completing these requirements and passing a comprehensive exam is the doctor granted a license to practice.

Many fire and emergency services organizations can improve their department's reputation and ability to gain a fair share of tax dollars by adopting minimum professional requirements for officers. Although this is a complex and controversial subject, the following are some examples of requirements that might be valuable:

- Minimum of 3 years' experience at each prerequisite level—fire fighter and fire company officer
- Certified as a Fire Officer III or IV

- Completed training and certified as a Fire Fighter II and Fire Instructor I
- An associate's degree for smaller departments (five or less stations) or a bachelor's degree for larger departments (from an accredited college or university recognized by the U.S. Department of Education)

Networking

Some more experienced chiefs have observed, "Whether elected by the department or appointed by the elected officials of your city, you, the new fire chief, are faced with some challenges. As a former chief put it, 'There's only one person who is more lonely than the greenest recruit on the department—the new chief'" (*Minnesota Fire Chief's Magazine,* 2002).

Chief Ron Coleman noted:

When that same fire chief who was a courageous fireground commander goes out on a limb to achieve the goal of making the community less vulnerable to attack from fire through sprinkler ordinances or by addressing other issues, they often are attacked from within their own ranks. There are people who jump at every opportunity to criticize the management of an organization. The person who was more responsible for the decision was the one who had to be confident that the decision was right. You can call that the loneliness of command, because that is what leadership is when we put it on the line. (Coleman, 2004)

The fire and rescue services are not alone in experiencing this problem; it has been around for many years, as expressed in the following quote:

Command is lonely. Harry Truman [President of the United States, 1945-1953] was right. Whether you're a CEO or the temporary head of a project team, the buck stops here. You can encourage participative management and bottom-up employee involvement but ultimately, the essence of leadership is the willingness to make the tough, unambiguous choices that will have an impact on the fate of the organization. I've seen too many non-leaders flinch from this responsibility. Even as you create an informal, open, collaborative corporate culture, prepare to be lonely. (Powell, 1996)

This points out the essence of the problem with getting fair, accurate, and unbiased advice and feedback on controversial administrative decisions: How does an administrator acquire the knowledge and advice needed to make an informative decision?

Other administrators or professional acquaintances may be able to help in evaluating and discussing changes, especially if these discussions would be better kept confidential in the preliminary assessment. While attending conferences, college classes, and National Fire Academy courses, chiefs can meet with professional peers and acquire e-mail addresses and phone numbers so they can stay in touch.

When administrators start thinking about changes for their agencies, they can ask for advice, thoughts, and information from trusted peers in other organizations. Because these peers are often hundreds of miles away, the advice can be unbiased. This is a method to validate a first impression on a proposed change without facing a full revolution by the members or elected/appointed officials.

Asking for advice, experiences, and information from others may be very troublesome for some chiefs. If supervisors, officers, or chiefs feel uncomfortable asking for help and advice, they must find a way to overcome this reluctance if they are to get unbiased private guidance from reliable expert sources. Finally, ask more than one person to make sure to get the entire picture.

To illustrate this process, looking at how a patient would deal with a life-threatening medical problem is helpful. The decision to undergo major medical surgery should not rest on the opinion of one surgeon; consulting with two or more doctors assures that all consequences will be revealed. To continue with this medical analogy, if faced with a recommended surgery, the patient should ask other doctors who are not surgeons for advice. Everyone has biases. This goes for temperament also; have a mix of traditionalists and risk takers who can be contacted for advice.

Finally, gather as many facts and data as possible. Look for others who have tried the same changes and gain from their experience. Study and analyze this information. This analysis may also disclose that the change cannot be justified. The formal process of research and analysis (policy analysis) for decision making will be discussed in detail in Chapter 11.

Modern Organizational Theory

Some argue that empowerment, as opposed to direct supervision and standardization, is the best way to coordinate the work of individuals who must constantly learn new skills and deal with changing environments. This is becoming the newest management technique to improve a business's service and competitiveness, and it works well in the entrepreneurial business world where a large part of the mission is to generate as much revenue as possible.

How would a fire and emergency services organization operate if it used empowerment and allowed each company officer to determine what actions the company would take at an emergency incident? This is precisely the reason why SOPs are needed to coordinate efforts at an emergency incident. Each company works as part of the overall team effort. Just as is the case in a team sport like football, if someone misses an assignment it generally hurts the entire team's effort.

Although empowerment theory may work very well in a corporate or industrial organization, it does not lend itself to the fire and emergency services organization at the company level. This is a good example of where popular contemporary management practices may not be appropriate without some modification for the fire and emergency services.

The emergency operations of a fire and emergency services organization run more smoothly if a combination of direct supervision and standardization are used. Employees or companies doing their own thing (freelancing) at the scene of an emergency cannot be tolerated. This is especially critical during the first few minutes after arriving on the scene. This is the time before the incident commander has arrived and has been able to evaluate and formulate a comprehensive strategy for the emergency. Consistent and reliable emergency operations at all times are very important, if not critical, to the successful and safe handling of emergency incidents.

In the administrative process, the chief would assign task groups of members and have them review national or other organizations' policies or strategies for emergency operations as the basis for their deliberations and final recommendations, thereby allowing members to prepare emergency SOPs and provide recommendations (**FIGURE 2-2**). The final decision and approval would still rest with the chief administrator.

FIGURE 2-2 Safety Committee Meeting (Alachua County Fire Rescue, Florida)

Staff-Line Distinctions

Most fire and emergency services organizations have a distinct dividing line between staff and line functions. This is especially true in larger departments where it is common to have specialized divisions of staff personnel such as fire prevention, training, dispatch, logistics, and human resources.

This traditional organization can be the cause of some serious conflicts and problems. First, the excitement and fun for fire fighters is working at emergency incidents, and the typical work schedule for career members tends to encourage many to stay in line functions. In addition, the crew can become almost a second family, which also increases the resistance to leaving shift work. As rookie fire fighters/paramedics, members become comfortable working their shift and look forward to the extended time off. In the common 24-hour shift, an employee is at work 1 day, then off for 2 or even 3 days. Many employees find time to work a second job during these days off.

In many departments, some of the brightest employees resist a transfer to a staff function or promotions above the rank of battalion or district chief because it usually means going to day work. These employees don't want to give up their part-time jobs and adjust their off-duty family and social commitments.

A New Paradigm for Staff Functions

As a solution to this problem, some departments have combined the staff–line jobs at the chief level. For example, a department may have a battalion chief in charge of each shift. This same battalion chief would also be assigned to supervise certain staff functions. During the weekday, the battalion chief would be the administrative head of prevention, but would be available to respond to major emergencies when needed.

The Madison (Wisconsin) Fire Department made this type of change in 1996. Partly due to a reduction in its budget, the number of division chiefs was reduced by combining staff and line functions. In a 1996 *Fire Engineering* article, Phillip Vorlander stated that "In most medium and large sized American fire departments, the assignment of fire service managers has not changed significantly since the days of horses and steam; duties are divided between classic line and staff functions." He went on to say that "While this system provided for effective emergency scene management, it fostered the creation of six mini-departments, each with its own cultures and allegiances. . . ." Vorlander then continued, ". . . shift commanders [line functions] often were alienated from the rest of the management group [staff functions] as a result of their involuntary exclusion from the decision-making process" (Vorlander, 1996, p. 131).

Each battalion chief now works two 24-hour shifts and six 8-hour shifts in a 2-week period, totaling 96 hours (a 48-hour work week). Between the 8-hour shifts and the 24-hour shifts, each battalion chief is able to cover 4 days each week at the staff supervisor position.

The greatest advantage of this type of staff–line consolidation for the Madison Fire Department is that:

> The availability of all staff officers to participate in the debate and consensus building process has markedly improved the quality of decisions. Awareness of all factors bearing on the decisions has also enhanced acceptance. Input into day-to-day decision-making has led to greater commitment to the management team. Also, the fact that chief officers don't have "ownership" of a particular battalion, division, or shift prevents issues such as performance, assignments, and so on from degenerating into turf controversies. (Vorlander, 1996, p. 131)

Another successful technique is to automatically transfer command officers every few years on a schedule that puts them into supervision of both staff and line divisions. This has the additional benefit of assuring that the next chief will have a well-rounded knowledge of the entire organization.

Supervision

Some public administrators would suggest that empowering supervisors would solve all management problems. Again, the fire and emergency services organization will not run at its best if chief officers negotiate the performance of each company. At emergency incidents, the fire and emergency services organization needs direct supervision and adherence to rules and regulations to operate efficiently and consistently. There are exceptions to this close supervision in some staff positions, however.

"An organization is in trouble if its official decision makers announce a policy, only to see it ignored or even subverted because those who are supposed to carry it out are battling with one another or pursuing goals of their own" (Garvey, 1997, p. 85). Because the typical fire and emergency services organization is geographically spread out, direct supervision is not possible in the classical meaning. A technique such as managing by walking around (MBWA) can be very effective in these situations if coupled with a strict enforcement of the rules and regulations.

MBWA is a technique used to increase communications at all levels of an organization. Generally, top administrators physically visit all areas of the organization. Most traditional organizations rely on the chain of command to both receive and send information from the different divisions and units.

MBWA helps to offset the communications problems that are inherent with large and dispersed organizations. Information about what's really happening gets distorted or, perhaps, purposely suppressed as it travels from bottom to top and top to bottom. Using MBWA, fire and emergency services chiefs can visit their stations and ask the personnel about a recently issued order or SOP. The chief will find out firsthand if the word is getting out. Very detailed observation at emergency incidents and a formal critique procedure with a written report to the chief can provide an accurate assessment and accountability of emergency operations.

Many MBWA administrators consider themselves facilitators and listen very carefully during their out-of-office visits for any signs that things are not going well. They will use their senior position to help remove any roadblocks so that a worker, unit, or division can accomplish their goals and mission. This can send a very powerful message to the members that the chief truly believes that their work is very important to the mission of the organization.

This MBWA technique can also be very effective when dealing with supervisors, elected officials, neighboring officials, and other peer department heads in government. This is a people skill that will help you gain support and cooperation. With this method, the chief visits and becomes acquainted personally and professionally with those influential people who can help. This should include people both below and above the chief in the chain-of-command, when possible.

An Example of MBWA

Some strong proponents of MBWA suggest that high-ranking administrators spend up to 60% of their time out of the office in the field. The following is a good example of the positive effect of using MBWA:

A safety chief for a large metropolitan fire department was in a staff vehicle traveling to a meeting and heard a dispatch for a townhouse fire over the radio. At this time, the department had just started a comprehensive safety program. General orders covering safety policies and SOPs had recently been signed and issued by the fire chief.

The safety chief was dispatched to only second alarm and higher emergency incidents, but he was in the neighborhood and decided to proceed in a non-emergency mode to observe the companies arrive and operate at this fire. About two blocks from the scene, the safety chief was passed by the first due engine. After it passed, he noticed a fire fighter standing on the back step. This was in violation of the department's safety general order for emergency vehicle operations.

After arriving on the scene, he continued to watch the fire fighters go to work at the townhouse, which had heavy smoke pouring from the front door and second floor windows. He then observed one fire fighter place a ladder to the front of the structure, climb the ladder, and start breaking out the windows from outside. This is a common ventilation procedure to remove heat and smoke from the structure on fire. However, this fire fighter only had a helmet on and no other protective clothing. This was another violation of safety rules.

The fire was quickly extinguished by the first alarm assignment. After the fire was extinguished, the safety chief continued his trip to the meeting without speaking to anyone on the scene. The battalion chief in charge of this incident was a personal friend and the friendship had started when the two chiefs were members of the same recruit school. It is not uncommon for officers to have to supervise co-workers who are close acquaintances. When the battalion chief

returned to his office, the safety chief called to advise him of the observations. Formal disciplinary actions could have been initiated, but that would have breached the normal chain of command, which is not the objective of MBWA. As the two spoke, the battalion chief immediately volunteered that the fire fighters had already admitted to the violations of the safety rules. The fire fighters had recognized the safety chief and knew that they had been observed. As a result of this confrontation, these individuals are committed to following the safety rules in the future. This is a preferred result of MBWA—to gain compliance with rules and regulations, while also demonstrating support for the worker's safety.

Bargaining Power

In most cases, when a new vision or policy is proposed, the chief must negotiate or convince others that the change is good. Most administrators would think that they have the advantage in bargaining because they are the recognized leader. However, bargaining power only starts with the formal position; it also includes professional reputation as well as knowledge and skill in debating.

There may be opponents to the new vision because some may lose privileges or simply because they resist any change. This does not mean that it is impossible to make policy changes, but the administrator must be prepared, informed, and patient. Have a plan to compensate losers (if possible) and make sure justification for the change is based on solid research, logic, and professional judgment.

Administrators who have a trustworthy and justifiable proposal may be able to negotiate an agreement with the members to implement the policy. For example, faced with irrefutable justification for the betterment of both the public and the individual, many labor organizations will cooperate with a negotiated agreement.

How would a person make another believe something in this type of negotiation? The answer depends on the question: Is it true? If administrators have the facts and can prove them, they have the bargaining and influence power.

Informal Organizations

Informal organizations can be used to help facilitate the negotiations and data gathering that must go along with the decision making and consensus building revolving around a proposed public policy objective. If not approached correctly, informal organizations can sometimes initiate and facilitate resistance to change:

Informal organizations emerge not only within and around formal structures but also as vehicles to facilitate contacts across the boundaries that divide different formal organizations from one another. In the modern administrative system, informal organizations embrace both private contributors and public officials. The flow of data [in informal organizations goes] both ways—from private citizens to public officials and also back from government administrators to citizens who are their clients or customers or the target of their policies. (Garvey, 1997, p. 238)

Administrators in government agencies or independent regulatory commissions should stay in contact with a number of individuals outside the official's formal organization, such as staff members of legislative committees, researchers from nearby think tanks, and representatives of lobbying firms and public interest advocacy groups. In addition, other government agency administrators and their staff need to be contacted on a regular basis to facilitate information flow and, when needed, support for policy changes.

An effective administrator must take full advantage of the informal organization to gather facts and acquire support for any policy changes. These are people skills that are needed to be successful in this process.

One important step is to identify and become acquainted with the community's power elite at the earliest opportunity. The power elite are a small number of highly influential citizens, and are rarely seen at public meetings. These are local people who have easy access to elected officials as a result of longstanding friendships, wealth, or influence at the state or federal level of government. Its members can either help with the adoption of a new policy or, with active opposition, doom the policy to failure. In most cases, it will be unusual for the power elite to get directly involved in the fire and emergency services policy decisions. For major policy proposals, however, such as the construction of a new fire station or a request for funding that would require an increase in taxes, the chief should consider approaching these individuals privately and explaining the request personally.

Although their help may not be needed in all cases, consider their potential influence during any attempt to change policy.

The following are examples of potential conflicts with the power elite:

- A new fire station on a vacant piece of property is proposed. The information about the neighboring properties contains the name of one of the power elite. However, the chief fails to make a personal contact with this person before announcing publicly that a site has been selected.
- A new physical fitness program is proposed that contains reasonable goals along with an implementation plan for those existing members who cannot comply immediately. However, the fire fighter's union was not consulted. The president of the union is a longtime resident of the jurisdiction and knows many of the power elite by first name.
- The chief has just finished a high-rise fire seminar at a national conference. Upon returning to the city, a reporter asks for any revelations on fire safety that were discussed at the conference. The chief announces to the reporter that he is now convinced that all existing high-rise buildings should be equipped with automatic sprinklers, and that a proposed local ordinance will soon be submitted to require it. What the chief is not aware of at that time is that the power elite own many of the existing high-rise buildings.

When faced with a conflict with the power elite, the chief is not always fighting a losing battle. Approach the potential change by slowly building a consensus among those people who will have an influence over the final approval.

In general, make arguments that are backed up by professional judgment and solid research (facts) for the best chance of receiving approval for a policy change request. If national consensus standards, such as those of the NFPA, can be used to support the request, it will be easier for the power elite, the elected officials, and the public to support the change because they will clearly understand the need and will shy away from any public debates.

There has been controversy within the fire service over the use of NFPA fire service standards. In some areas, local fire and emergency services officials want the freedom to provide the level of public fire protection that is determined necessary by the local community. In many cases, the citizens did no research or planning to determine the local level of protection or quality of service (e.g., number of stations, companies, and staffing). Whatever resources and types of organizations exist at the local level are the result of a very complex and independent evolutionary process, not a systematic planning effort.

Fire and emergency services organizations are in the process of entering a new era. In the emergency medical services field, responders are now certified to national standards for emergency medical technician and paramedic. In the fire services arena, national standards for training and safety have been available for a relatively short period of time. These national standards can be an opportunity to gain approval of policy changes.

For example, the chief may propose a new program to certify all officers to NFPA 1021, *Standard for Fire Officer Professional Qualifications*. This type of program may require training, incentive pay, and overtime pay to cover classroom attendance (although this may not be necessary if training and education are a requirement for promotion). The selling point for this proposal is that, when completed, the city will have a more professional, competent fire department. This is the kind of proposal that the municipal administration, the public, and elected officials can easily understand.

Remember that the power elite and special interest groups (such as fire fighters' unions and volunteer fire associations) can have a big influence over elected officials. Approach these groups and their officers at the earliest opportunity to gain their support. Again, it is not always absolutely necessary to gain their support to be victorious, but in many cases their support can make the process smoother, resulting in a better chance of success.

Presidential Power

What happens when a person is promoted to chief? Does being the chief guarantee that all the members will follow all of your orders? A fire and emergency services administrator, when managing and leading change, would do well to study how the most powerful administrator in the world uses position and power to make changes.

In 1952, Dwight D. Eisenhower, who had been a great army general during World War II, was elected president. The outgoing president had this to say about Eisenhower: "He'll sit here," Truman would remark, "and he'll say, Do this! Do that! And nothing

will happen. Poor Ike—it won't be a bit like the Army. He'll find it very frustrating." "In these words of a President, spoken on the job, one finds the essence of the problem now before us: 'powers' are no guarantee of power; clerkship is no guarantee of leadership." (Neustadt, 1990, pp. 9–10)

The easiest form of power is the direct order or command. The following criteria are necessary for a command to be followed without question:

- The reason for involvement is instantly recognizable.
- The orders are widely publicized.
- The words are clearly understood.
- The supervisors and members who receive the order have control of everything needed to comply.
- There is no doubt about the authority to issue the command.

It is rare that orders on the emergency scene are not carried out verbatim. Because the fire and emergency services are quasi-military organizations, it is well understood by all members that they must obey the orders of a higher-ranking officer. Officers are trained and educated to follow orders even if the justification is not provided or not understood. Individual officers and their company assignment will fit into an overall plan that may not be clear to a specific unit, but fits into an overall strategy to mitigate the emergency. Therefore, orders must be obeyed without question, unless it is clearly an unsafe directive.

In nonemergency situations, the ability to effectively administer the fire and emergency services organization will ultimately depend on whether the chief can persuade or convince others of what ought to be done for their own good or for the betterment of their service to the public. In addition, follow-up (MBWA), deadlines, and delegation all play a very important role in everyday managing and change projects. This may include occurrences when discipline is necessary. The following are all methods of persuading the members to comply with the chief's leadership.

Order Compliance

A good example of how employees may have their own agendas that may keep them from complying with an order became apparent during President John F. Kennedy's administration. In 1962, the President became aware that the Soviet Union was constructing nuclear missile facilities in Cuba. Because Cuba was so close to the mainland of the United States, these missiles could be launched and hit targets in the United States in an alarmingly short time.

As it became clear that his negotiations with the Soviets were not making any progress, President Kennedy ordered a blockade around Cuba. A blockade can be construed as an act of war. This was a very serious crisis that brought this country to the brink of nuclear war with the Soviet Union.

During the negotiations that followed, an offer was received from the Soviet Union that if the United States would remove nuclear missiles from Turkey, the Soviet Union would remove its missiles from Cuba. At the time, Kennedy remembered that he had previously given a presidential order for the U.S. military to remove its missiles from Turkey. Kennedy had received expert advice that the missiles were not necessary for the defense of the United States and could safely be removed.

The military officials had deliberately failed to comply with this presidential order; military officials who are trained to obey all orders by superiors did not carry out an order by the most powerful person in the United States, the commander-in-chief. The military officers believed that these missiles were critical to the defense of our country. Fortunately, this provided Kennedy with a bargaining token to make a deal with the Soviet Union. The United States still had the nuclear missiles in Turkey that Kennedy then used in negotiations to trade for the removal of the Soviet missiles from Cuba.

However, after this experience, Kennedy set up a system to follow up and collect information about what was really going on in the many levels of the government. Not able to MBWA with such a large organization, he established direct contacts with friends and trusted confidantes throughout the government.

Accurate feedback can be hampered either by subordinates following their own agendas or by high-ranking government officials not wanting to give the boss any bad news. This type of behavior by subordinates can be prevalent in departments where the administrator has a lot of influence over who is promoted. Many subordinates see their chances of being promoted as being based on their ability to both do a great job and keep the boss happy. Colloquially, this is called surrounding yourself with yes people. In Kennedy's case, he did not want to be blindsided again, and made extensive changes to provide impartial avenues by which information could reach him.

A fire and emergency services administrator should follow this example. The ability to recognize a problem and take action to make sure it does not happen again is a great attribute for an administrator. Similar to the president's situation, some fire departments are so large that the chief cannot visit each station in a year. However, this information problem can happen even in the smallest organization. The chief should establish an informal information gathering system made up of friends and confidantes throughout the organization.

Professional Reputation

Once you have the formal position of chief, the most important factor that will enhance your persuasive ability is your professional reputation. If the chief has the reputation of not following up on assignments, then when a request for members to comply with one of the department's orders is issued, either verbally or in writing, many will not comply. This is especially true if the members do not agree with the change or request.

How can a chief build an outstanding professional reputation? The following is a list of some suggested items that can help build an outstanding personal and professional reputation:

- Build a solid experience base, starting at the lowest level and moving up through the ranks. This does not mean that chiefs will still be competent in basic skills, but that they will have been competent in the past. In small organizations, up to about three stations, the chief may have to help out and still should be skilled in basic emergency tasks.
- Take advantage of every opportunity to learn the skills and knowledge of the job, through both hands-on experience and training.
- Keep in the best physical shape possible; this includes strength, aerobic fitness, and body weight within recommended limits. Even if the chief no longer has to perform physical skills (see the first item in this list) this is a great idea for the chief who believes in leading by example.
- Acquire a formal education, especially through programs that lead to an associate's or bachelor's degree in fire science, administration, management, and other job-related areas.
- Stay up to date by attending seminars and conferences and reading fire and emergency services, government, and management periodicals (a course in speed reading is very helpful). Some chiefs in larger departments have staff review and provide a synopsis of any pertinent articles.
- Read all local newspapers and influential organizations' newsletters. Be familiar with

all controversial issues in the community, not just fire and emergency services.

- Visit other fire and emergency services organizations. There are good ideas everywhere.
- Have high personal and professional standards for ethics and morality.
- If appointed to a top position, choose your initial proposed changes carefully. Choose only those changes that are certain to succeed. It is very important to start out as a winner.
- Last, but perhaps most important, lead by example.

Feedback from Staff

There are situations for which the decision maker does not know all the facts or circumstances and consequently makes bad decisions. This is sometimes called imperfect information. This may happen when the top staff members are reluctant to make an argument that opposes what the chief would like to do.

When selecting staff, administrators should look for those who would be loyal. The most important reason to look for loyalty is that it allows you to communicate honestly with staff and have the conversations remains private. In many of these conversations, the discussions, thoughts, and proposals may be unsuitable for the public, other members, or elected officials. If these preliminary discussions find their way to others, these discussions may cause unnecessary hysteria and controversy.

For example, a serious problem with members having heart attacks has been identified in the department. When discussing this problem, the chief and staff privately talk about a new medical and physical fitness standard for new and incumbent members. One staff member makes the observation that people who run marathons appear to be in great physical shape and would not be as likely to have a heart attack; if they did have a heart attack they would probably have a better chance of surviving. This is a true statement. Can you imagine the reaction of existing personnel to the possibility that there would be a requirement to finish a marathon to continue active emergency operations participation? At this point the smart chief would ask the staff person to privately research the issue and provide an overview of all options. It would become clear from the research that

fire fighting, although very strenuous physically, does not require members to be at the physical fitness level of a marathoner.

There is a difference between loyalty and blind loyalty. The administrator must be careful to always solicit opinions in an open format. Preliminary discussions with staff may be held in meetings behind closed doors to facilitate an open, honest dialogue. Once the administrator has chosen a particular policy, staff would be expected to support the change. Never equate a privately expressed negative or opposing view as a sign of a person's lack of loyalty. In fact, this person may be giving the chief the advice needed to stay away from a bad decision.

When the subject is one of major importance to the organization, it is best to empower a task force, containing members representing diverse opinions, to study and make recommendations. Choosing members from all levels, from experienced fire fighters to officers, provides accessibility to opinions and advice from all levels. This truly opens communications and prevents any information obstructions that are common in middle- and upper-level management.

How do administrators get their personal staff to be completely open and present all their facts, opinions, and knowledge? First, chiefs should encourage open and honest discussions and explain that it will help make fair and superior policy decisions. Also, when a staff member brings bad news, be careful not to react angrily. An angry reaction can discourage the staff member from bringing any negative information in the future.

Finally, chiefs should acquire information through their own personal research, by contacting numerous informal sources and special interest groups, such as unions.

Visit stations often and float ideas for informal discussion. Be careful, however, because there are always some members who may not be able to differentiate a brainstorming session from a formal announcement of a new policy. Develop a keen listening ability. During actual implementation, confirm that all the details to fully implement the decisions have been completed. Be your own intelligence officer. This is all part of a successful MBWA program.

In some cases, the chief has the authority to order a change but members may make this change in name only, and not in action. Many departments will report to their peers and their community that they are progressive and modern from what they say, not what they do. The chiefs of all departments need to use

MBWA along with accountability to find out what is really going on in their departments.

The Ultimate Power

Do not underestimate the power that the chief administrator has to make change. In most cases, this power goes beyond simply giving orders, which may or may not be followed.

This is a very simple concept. When administrators speak and initiate a new project or change, the attention of the organization and its members will be immediately focused on the proposed change.

This brings up the chief's ability to set the agenda for change in the organization. Generally, the majority of changes come about when the chief initiates the process. This takes real courage and detailed preparation in most cases. The exceptions are changes that are forced from the outside by a court order, a new state or federal regulation, union pressure, or elected officials.

The chief's ability to get the administrative job done, identify problems, and, when necessary, make changes in the organization makes up a dynamic process. To comprehend these dynamics, an understanding of the chief's power and its interdependence on professional capabilities and reputation, along with the ability to influence others, is critical. This is especially true when change must be implemented. The following checklist should help you determine if you, as chief, are ready to manage and lead the march to initiate change:

- How is your professional reputation?
- Are you knowledgeable about all aspects of emergency operations?
- Do you have the appropriate formal education?
- Can you lead by example?
- Can you gather all the facts you need to make an informed decision?
- Do you really know what is going on in your department and community?
- Can you schedule and set an agenda for change that is reasonable and evenly paced?
- Will the change improve the service to the public (customer)?
- Will you take risks for the sake of making real progress in improving service to the public?

CASE STUDY: An Example of Policy Analysis

A voluntary physical fitness program increases the average physical fitness of fire fighters. It is well recognized that firefighting requires a high level of strength and endurance. Therefore, the assumption is that the public would be better served by fire fighters who can perform at more advanced physical levels.

Firefighting is a team activity, so the weakest link will limit the entire company or team. When a two-person team enters a structural fire using SCBA and one individual is in good physical fitness and the other is not, the total team effort will be limited by the less fit person. A person in good aerobic condition can have a useful work time with the SCBA of around 20-30 minutes, whereas the unfit person may not be able to stay more than 10 minutes before their low air alarm sounds. Therefore, the team of fire fighters will be limited to 10 minutes because when one fire fighter leaves, the other must also leave for safety reasons.

Conclusion: A mandatory physical fitness program is needed.

Discussion Questions

1. Make a list of both management and leadership goals that would have to be accomplished to implement the conclusion from the above case study.
2. Do you have a management or leadership preference? How would this affect your actions necessary to complete the change process? Give several examples.
3. If you are stronger in one preference, describe how you would select and incorporate people with the other preference into your implementation process.

References

Coleman, Ronny. "Chief's clipboard." *Fire Chief.* May 1, 2004, p. 5.

Fire Chief. "What's new on the safety scene." April 2001, pp. 50–60.

Garvey, Gerald. *Public Administration: The Profession and the Practice.* New York: St. Martin's Press, 1997.

Minnesota Fire Chief's Magazine. "The Minnesota Fire Chief's Responsibilities." January 2002, p. 3.

Neustadt, Richard E. *Presidential Power: The Politics of Leadership.* New York: John Wiley & Sons, 1990.

NFPA. *Fire Protection Handbook,* 19th ed. Quincy, MA: NFPA, 2003.

NFPA. About Risk Watch. http://www.nfpa.org/riskwatch/about.html (accessed July 22, 2005), 2004.

Powell, Colin. *Quotations from General Colin Powell: A Leadership Primer—18 Lessons from a Very Successful American Leader.* http://www.usna.edu/JBHO/sea_stories/quotations_from_general_colin_powell.htm (accessed May 17, 2003), 1996.

USFA. *America Burning: The Report of the National Commission of Fire Prevention and Control.* http://www.usfa.fema.gov/usfapubs/pubs_main.cfm, 1973.

Vorlander, Phillip. "An innovative approach to fire department command staffing." *Fire Engineering.* August 1996, p. 131.

Wilmoth, Janet. *Fire Chief Command Post: From the Editor.* http://enews.primediabusiness.com/enews/firechief/v/137 (access July 2, 2005), 2004.

3

Leading Change

Knowledge Objectives

- Comprehend change and its impact on contemporary organizations.
- Understand the dangers and roadblocks that can defeat efforts to accomplish change.
- Recognize the influence the chief administrator has on the likelihood of change success.
- Examine an eight-step process for change management.
- Comprehend the outside pressures on potential changes from members, municipal administrators, the public, the press, elected officials, and officials of special interest groups.
- Examine the internal and external forces that create demand for changes.

What Is Change?

Change happens continuously in the private lives of all people, as well as in businesses and governmental organizations. Change is particularly prevalent in business, and it is absolutely essential for a business to survive and to prosper. And this change is not going to slow down anytime soon. Worldwide competition will probably speed up change even more in the future. The expectation is that to stay competitive, businesses must constantly change to meet customers' expectations.

Change has been called many names, such as restructuring, re-engineering, transforming, acquisitions, mergers, innovating, modernizing, downsizing, rightsizing, quality programs, and cultural renewal. Rightsizing (sometimes called re-engineering or downsizing) in the business world has been very disappointing at times, and the carnage has been appalling, with wasted resources and burned-out, scared, or frustrated employees. The major emphasis of this management technique is to do more with fewer employees. This leaves those lucky ones still working for the company always looking over their shoulders for the next wave of rightsizing. In addition, employees are expected to produce more and work longer hours for the same pay.

In many cases the changes do not become permanent. The organization's members resist change and are able to revert back to old practices when the current leader leaves. Change is very difficult to institutionalize, and this is the challenge of leadership.

In the business world, strong macroeconomic forces are at work, and these forces only grow stronger over time. As a result, more and more organizations will be pushed to reduce costs, improve the quality of products and services, locate new opportunities for growth, and increase productivity. A good example of this in the fire and emergency services is the growing number of fire departments providing emergency medical services.

Doomed Change Initiatives

Several errors can doom change. The first is allowing complacency. This is a result of not properly influencing or educating employees and managers of the urgency to make the change. In the private sector, it can be easy to recognize this error when the company starts to experience a dire financial situation such as not meeting sales goals or profits dropping below expectations. Public safety organizations can continue the status quo for many years without any outward indication that service delivery is below average or not up to the latest state-of-the-art equipment, training, strategy, tactics, or management practices.

In these cases, an administrator has to make the first call and stand up for proposing changes to correct shortfalls in services. In many cases, the public, elected officials, labor organizations, and fellow managers may not be aware that changes are needed to improve the quality or efficiency of service.

Another error that can doom change is overestimating how much an administrator can force big changes on an organization. It is easy for a chief to underestimate how hard it is to drive people out of their comfort zones. In addition, chiefs often do not recognize how their own actions can inadvertently reinforce the status quo. Leading by example is a very strong influence on members, both positive and negative. Finally, many administrators lack patience.

Vision

The role *vision* plays is one of the more important elements of a successful change. Vision plays a key role in change by helping to direct, align, and inspire actions of large numbers of people. A vision is like a goal. Administrators must know where they plan to end up before starting the journey.

Once the administrator has a vision, it is extremely important that it be communicated to the organization's members, the public, and appointed/elected officials. Communication can be accomplished with both words and actions. The latter is generally the more powerful of the two.

There will always be some people who do not buy into the change the first time they hear about the new idea, and some will take more convincing than others before they become believers. Others will never be convinced, but don't let this minority resistance stop the pursuit. Be patient and persistent.

It is common to find that a small minority of members and supervisors will actively resist the new vision. Wonderful incentives or even the greatest leadership will not persuade them. Therefore, a disciplinary process should be instituted; its final stage may have to be the separation of the person (or persons) from the organization, although this would be extremely rare. Faced with this type of consequence, it would be very unusual to find a member who would not comply when given a reasonable time to conform.

To either rely on voluntary compliance or institute rules that are not enforced is to invite failure. This will be perceived as a lack of administration support. Even in the largest organization, however, disciplinary action may only be necessary for a few members. The word will spread quickly that the chief means business.

A Safety Example

A fire rescue department issued driving regulations that contained a provision that all emergency response vehicles must come to a complete stop at all red traffic signals and stop signs (as per National Fire Protection Association [NFPA] 1500) during emergency responses. Each paramedic unit was equipped with an automatic speed recorder. A crash occurred one day when a paramedic unit failed to stop at a red traffic signal, striking a car that was proceeding through a green light, killing the driver of the car.

The "stop at all traffic control signs and signals" regulations had been issued about 1 year prior, but had never really been enforced. It was common for many drivers to proceed through the stop signs and signals without coming to a complete stop. When these circumstances were relayed to the jurisdiction's lawyer, he recommended deleting the driving regulation, stating that if the department was not going to enforce the regulation, it would make the jurisdiction more liable. He felt it was better to delete the safety regulation than attempt to enforce the rule. Whenever smart and well-intentioned people avoid confronting obstacles, they disenfranchise employees, supervisors, and the public we serve, and they undermine change.

The investigating police officer did not charge the driver of the paramedic unit. This was a standard professional courtesy at the time, but more recent obvious violations have resulted in the fire/rescue member being charged with a moving violation. An out of court settlement was achieved with the relatives who had brought a wrongful death lawsuit against the jurisdiction. The chief was able to persuade the attorneys that the safety regulation was imperative and a plan was implemented to enforce the rule. This is the hard way to learn a lesson—the loss of a life and a settlement of many hundreds of thousands of dollars.

Change Created by Standard Operating Procedures

Many times in the fire and emergency services organization, standard operating procedures (SOPs) are the vehicle used to create change. The creation of SOPs can cause the implementation of new methods that are safer or more efficient than current procedures. However, some legal experts have advised that departments should not have SOPs because this makes them *liable*.

Many of these SOPs are in response to safety issues and the need to provide better service to the public. These are items that can cause liability exposure if not addressed by the department. For example, if there is a failure to provide for the safety of the members or to give adequate service to the public as measured by nationally accepted professional standards, there will be an increased liability exposure. If the department fails to enforce the SOPs, then there may be a doubling of the exposure to possible legal action.

It is important to remind yourself that real transformation takes time. Some major changes may take generations. Patience is needed to complete the transformation as guided by the newly created SOPs.

For example, in the case of physical fitness standards, it may be reasonable to use a 6-year implementation period for existing personnel. It took many years for these fire fighters to become out of shape, so it would seem fair to allow them a reasonable amount of time to achieve minimum physical fitness levels. However, each year should require employees to reach a specific short-term goal.

"Most people won't go on the long march unless they see compelling evidence within six to eighteen months that the journey is producing expected results. Without short-term wins, too many employees give up or actively join the resistance" (Kotter, 1996, p. 11). Some of these short-term wins can be recognition, promotions, or money. In addition, disciplinary actions, following the standard local procedures, should be taken at these short-term benchmarks for those who fail to meet them. For example, the first notice could be verbal to the employee and a written note inserted into their personnel file; second notice would be written to the employee specifying the infraction, time for compliance, and consequences of noncompliance; the third and final step may be termination.

Anchoring change into the organization's culture should be a primary goal. Permanent change does not occur until the overwhelming majority of the members believe in the new vision. All members of the administration must continuously lead by example.

Securing change also requires that sufficient time be taken to ensure that the next generation of officers really does embody the new vision. Promotion criteria should be revised to conform to the new changes or transformation will rarely last.

For example, if the new vision is to upgrade the formal education of the department's officers over the next few years, the best way to do this is to either require a minimum education level for promotion or provide additional points to the score on the next exam for officer candidates who meet the education requirements. This is another change that, if implemented slowly, allowing all present employees the opportunity to go back to school and acquire the required education, can be accepted more easily and perceived as fair. This may take many years, but the change will encounter less resistance and will have a better possibility of survival.

Creating Change

Change or transformation has been very slow in the past for the typical fire and emergency services organization. Tradition has been the norm, and it is a very comfortable place for most members to reside in. Any possible change or person who rocks the boat with new ideas becomes a target for the majority of employees who resist change. Change is a direct threat to their comfort zone.

When change is forced on people, which administrators sometimes do, members may be able to find a million clever ways to undermine the change. In other words, authority by itself cannot ensure that changes will be permanent. Leadership is a set of processes that creates progressive organizations or allows them to adapt to significantly changing circumstances and environments. This is not an everyday occurrence.

This resistance to change becomes a double handicap when combined with traditional managers who have not been trained in leadership skills. Although changes can be implemented by issuing a new SOP or order, to be completely successful at making the change last in the future, change needs to be applied using a plan.

Terms such as *thinking out of the box, future vision,* and *brainstorming* all speak to methods used to identify changes. Once change is identified, leadership must take over to produce permanent change that can be successful in transforming an organization to better service.

The following eight steps will help develop a successful plan:

- *Step 1:* Identify the problem and create a sense of urgency.
- *Step 2:* Create a guiding coalition.
- *Step 3:* Develop a vision and strategy.
- *Step 4:* Communicate the change vision.
- *Step 5:* Jump over barriers.
- *Step 6:* Create short-term wins.
- *Step 7:* Be prepared for resistance.
- *Step 8:* Finalize and institutionalize the change.

Step 1: Identify the Problem and Create a Sense of Urgency

About 10%–15% of the members of the organization must commit to the change before it can be successful. This is necessary because some changes contain sacrifices. Sacrifices are generally more common in the business world than in government, because most of the benefits of employment in the public sector are somewhat guaranteed, but one major exception to this would be any changes that threaten jobs or the membership of volunteers.

For example, many fire departments have added EMS to their scope of services. This change requires most of the members of these fire and emergency services organizations to achieve some new level of training, such as First Responder, EMT-Basic, or EMT-Paramedic certification. Although each situation is unique, fire and emergency services organizations have offered benefits such as incentive pay and paid time off to attend classes to offset the sacrifices that would have been necessary had the organization simply mandated the new levels of job performance.

In the case of many volunteer organizations, these sacrifices are not offset by any monetary incentives. This is a good example of leadership creating the urgency for change despite substantial sacrifices. Because volunteers are very dedicated to their organization and the public they serve, they generally attempt to adhere to the requests of their administrators. However, it appears that the new EMS demands on volunteer fire and emergency services organizations have resulted in a reduction of members as a result of the additional time needed for increased call volume

and the additional effort and time needed to acquire new training and skills. This has had an adverse impact on many volunteer organizations, and is an example of an unintended consequence of change.

The fear of the unknown tends to magnify the doubts of the organization's members about the potential sacrifice. To create a *sense of urgency*, the cooperation and willingness of a small group of core members is needed to support the change initiative. In a public service organization, the support of elected and appointed officials is also critical; the same 10%–15% support rule also applies to them. The chief administrator will need champions both above and below them in the organization to support the change.

No matter how hard senior-level administration pushes, if others don't feel the same sense of urgency, the momentum for change will probably die far short of the finish line. In a paramilitary organization it is always possible to force change by issuing a new SOP, but if the SOP is not based on some credible justification supported by a guiding coalition, then the change will only be temporary until the next administrator takes charge.

However, mandated change is always possible when implemented with some type of governmental regulatory power. Also, in some organizations it may be possible to gather the support of a core group simply because of the members' loyalty to either the organization or a charismatic leader. Avoid falling into this trap of support by blind loyalty; make sure the justification is based on facts.

Remember to ensure that a credible analysis of the requested change has taken place (see Chapter 11). This can be a very complicated process. Be cautious in making the choice to implement change. Wrongly based changes do not withstand the test of time.

Sources of Complacency To create a sense of urgency, several very real barriers that are part of the everyday workings of many organizations, including fire and emergency services organizations, must be overcome. Think about the last time that you attended a management, union, or volunteer association meeting. How much of the formal business deliberations discussed any change initiatives?

In most cases, no highly visible crisis existed at the time the meeting was held. In the case of a fire and emergency services organization, some highly visible crises could include cuts in the budget that may cause salary or benefit reductions, layoffs, or the closing of fire stations or companies; lack of funding to replace old and/or unreliable fire or rescue apparatus; or the threat to the membership or existence of a volunteer organization. Major changes have occurred when career fire fighters have accepted reductions in salary and benefits when convinced that the alternative would have been layoffs of fellow fire fighters. The urgency level was high for a clearly visible, credible crisis.

Because most fire and emergency services organizations receive funding from tax revenues, which generally are fairly constant and reliable from year to year, it is common for the following subliminal message to be present: We are winners and must be doing something right. So relax and enjoy the status quo. This is especially true in fire and emergency services organizations because these agencies do not depend on making a profit from products or services. If a business were losing profit and customers, there would be an outcry by the stockholders for change.

Normally, when fire and emergency services improve their product it is through expansion of services, new or improved equipment, or better-trained and competent members. Public or elected officials will not have a good understanding of the methods, justifications, or consequences of this type of change. In fact, it is possible in the fire and emergency services business to supply very poor service and not have it noticed by any of the customers. Remember that fire and emergency services have a monopoly and there is no direct competition, which also tends to eliminate a source for judging whether or not the organization is providing world-class professional service.

In the business world, where there is competition and free entry into the market, competitors are very quick to point out their better service. Imagine, for a minute, a town that has two competitive fire and emergency services organizations, company A and company B. The revenue for these companies would be the result of fees collected from customers who called them for emergency service.

The chief of company B notices that company A has made some changes. Company A has increased the staffing on all its companies to four personnel and is advertising on TV and in the local papers that it can provide better service by complying with nationally recognized professional and safety standards. Company A's ads characterize its new service to be World Class Quality. If the citizens had a choice (and no loyalty issues), whom would they call if their house or family needed emergency service? In reality, many fire and emergency services providers do not want to be compared to other similar organizations or professional standards, because they may not measure up.

Many of the situations that a business leader can use to create a sense of urgency are not available to the fire and emergency services organization. This makes the leader's job more difficult, but not impossible. The chief may have to look very carefully, and have the patience to wait for the opportunity to make a change. "A good rule of thumb in a major change effort is: Never underestimate the magnitude of the forces that reinforce complacency and that help maintain the status quo" (Kotter, 1996, p. 42).

Increasing the Urgency Level Traditionally, the fire and emergency services have used *crisis management* to achieve change. Whether the push for change came from outside or inside, change would not have occurred if it had not been for a major catastrophe that occurred either locally or nationally. Visible crises are enormously helpful in catching people's attention and increasing the level of urgency. Conducting business as usual is very difficult if the building is on fire or there is a firefighter death.

In the public arena, it pays to be prepared for these tragic opportunities. Have a plan and justification prepared. These crises are very predictable, but some may be highly unlikely locally. For example, in a nearby jurisdiction a fire-fighter death occurred or a tragic incident was reported in the media. These are opportunities to pump up the urgency for change. It is fairly common in the fire and emergency services business to respond to many of these incidents with new fire prevention code initiatives, especially at the national level.

For example, many changes were realized in Florida as a result of the notorious 1992 Hurricane Andrew and the 1998 wildfires, such as the purchasing of new firefighting and emergency equipment, the hiring of additional personnel, increased training for fire fighters, and the creation of multi-agency statewide mutual aid agreements.

In one case, the State of Florida's fire service identified a major problem in the ability of mutual aid fire companies to communicate by radio during several hurricanes and major wildfire incidents. As a result of these catastrophic tragedies, the state spent close to $1 million on portable radio communications equipment that can be dispatched to the scene of a major emergency incident. This equipment provides compatible radio communications using portable handheld radios and a portable base station with a 100-foot crank-up tower (**FIGURE 3-1**). There are seven of these radio communications systems in the state and, through the incident management system, each can coordinate over 100 fire rescue companies. The sense of urgency was clearly evident to the elected officials and the fire rescue service.

As a result of the unprecedented terrorist attacks on the United States on September 11, 2001, a major change is taking place in the federal government. Among many changes, federal grants that were slated to be eliminated from the federal budget are being funded again. This change did not happen in a vacuum, but was the result of many fire service administrators who pulled out their visions and collectively applied pressure on their federal legislators to provide help. It was agreed upon by all the administrators that more resources were needed for first responders at all emergencies, both everyday and catastrophic. The need was never so clear and the urgency was never higher.

In an increasingly fast-moving world, waiting for a fire to break out to initiate change is a dubious strategy. If possible, attempt to increase the urgency level prior to a major emergency.

Raising the Urgency Level in Good Times Raising the level of urgency when there are no crises is a classic problem for the fire and emergency services organization. As discussed previously, because of the nature of a public service organization, there is no competition. If the department needs a new station, equipment, or anything that will raise costs, there may be an uphill battle to gain approval for the funding. But there is hope. Catching the public's attention during good times is far from easy, but it is possible.

FIGURE 3-1 Radio Communication Tower (Hurricane Katrina, Mississippi, 2005)

The NFPA fire service standards are one good source for creating the critical urgency level needed for change. Many fire and emergency services organizations use them routinely. They are the best professional standards available.

Some organizations use only part of these standards to fit their specific change issues. The purchase of protective clothing and fire apparatus are commonly supported by pointing out that the department would like to comply with the safety standard NFPA 1500, *Standard on Fire Department Occupational Safety and Health Program*. In many cases the department doesn't attempt complete compliance because several of the items may require major change and sacrifice by members.

The items that are typically left out deal with issues such as training, incident command, staffing levels, and medical and physical fitness standards. These are admittedly aspects of change that can be costly and may be difficult to implement. But, if the vision of your fire and emergency services organization is to be one of the *best of the best*, the chief must tackle these more controversial changes. Most of these changes can be phased in over time (as noted in the NFPA 1500 standard), which will drastically reduce the potential impact and resistance.

Creating a Crisis Progressive leaders often create artificial crises rather than waiting for something to happen. It can be better to create the problem yourself. Better still, if at all possible, help people see the opportunities or the crisis-like nature of the situation without experiencing real losses. The type of losses that are evident in a fire and emergency services organization can be more public than the financial losses experienced in a business enterprise, such as a story on the front page of the local newspaper stating, "Fire Department Fails to Rescue Two Trapped Children in Tragic Fire!" Major property damage or fire-fighter deaths may also be a source of losses.

What kinds of changes would prevent this type of notoriety? For the sake of academic brainstorming, consider the following possibilities:

1. Fire and emergency services companies arrive on the scene with inadequate staffing.
2. Members of the department are not trained to nationally recognized professional standards.
3. Members are not capable of performing all the functions of firefighting because of medical or physical limitations.

4. The fire and emergency services company's response times are unpredictable and unreliable because of a lack of on-duty staffing.

Although, at this time, there has not been any case of the above-mentioned examples being used in lawsuits, this does not ensure against any future action. In the future, new NFPA standards will be created and existing standards are under continuous periodic revision. Remember that NFPA 1710, *Standard for the Organization and Deployment of Fire Suppression Operations, Emergency Medical Operations, and Special Operations to the Public by Career Fire Departments* is very new, and legal actions in the future will assuredly revolve around this standard.

To create a crisis, the chief may use the NFPA standards as justification to argue that with improvements in equipment, staffing, training, or procedures, the department will be able to provide better emergency service and be less liable. This is not the typical crisis that has to be solved in the next 5 minutes, as would be the case for an emergency in progress, so elected officials and the public need to become convinced that change must take place to solve a problem. In addition, this is the morally correct course of action that will provide the residents with the best public fire protection while also increasing the safety of fire fighters. This means formally and publicly making a proposal for change.

> Perpetual optimism is a force multiplier. The ripple effect of a leader's enthusiasm and optimism is awesome. So is the impact of cynicism and pessimism. Leaders who whine and blame engender those same behaviors among their colleagues. I am not talking about stoically accepting organizational stupidity and performance incompetence with a "what, me worry?" smile. I am talking about a gung ho attitude that says, "we can change things here, we can achieve awesome goals, we can be the best." Spare me the grim litany of the "realist"; give me the unrealistic aspirations of the optimist any day. (Powell, 1996)

Caution Caution Caution If you are presently in a role of middle or lower-level management and have a vision that will require changes to accomplish, keep these ideas to yourself or to those associates and friends who will not be threatened by the thought of change. Be aware that many people dislike even the simplest change, much less change that may require some sacrifice.

For potential future leaders to be effective at making change and contributing to the betterment of the profession, they must get promoted or elected (common in many volunteer departments) into a senior-level officer position before having the influence needed to make real change. If their ideas threaten fellow members or superior officers, they may be overlooked for promotions.

This does not mean compromising personal professional standards. For example, a company officer strongly believes that the organization could do a better job if members did more in-service training. The officer may not want to preach this belief to everyone, but may increase the training of his or her company.

"If everyone in senior management is a cautious manager committed to the status quo, a brave revolutionary down below will always fail" (Kotter, 1996, p. 48). Remember, there are organizations in which the entire top administration is against change. A future leader must be careful to seek out those who are open to change for support.

This points out another major difference between the business community and the typical fire and emergency services organization: It is not the norm to be able to switch agencies or employers when in middle management positions. If prospective leaders are easily frustrated and not patient, they may be in the wrong profession. It can take many years to climb up the promotion ladder to get to a place where you can effect real change.

In many cases, the only truly effective place is at the top. Even the chief may be limited by elected officials, city managers, or, for volunteer organizations, those who elected the chief. And it is not unusual for employee organizations to have influence. However, by following the change step process, administrators will have their best chance of leading an organization through significant change.

Step 2: Create a Guiding Coalition

As the top leader, why can't the chief just dictate change and have everyone comply? The answer to this question is found in the nature of organizations and the people affected, which can cause significant change to be resisted at many levels.

One of the most important necessities for successful change is a strong guiding coalition—one with the right composition, level of trust, and shared vision. Typically, a committee is formed that has the responsibility to recommend implementation and monitor progress. These must be people who are committed to the change.

How to Choose Committee Members One contemporary issue involving change in many fire and emergency services organizations is the implementation of a physical fitness program. The facts that support this type of program are very compelling and based on solid research. In addition, from a common sense perspective, having members in good physical fitness is good for their health and provides an emergency services worker who can perform at greater levels of strength and endurance (**FIGURE 3-2**).

To implement this change, the chief would want to identify those people who could be appointed to the physical fitness committee (guiding coalition). One technique that can be successful in identifying committee members is to implement a pilot program and ask for volunteers.

There may be opposition to the makeup of the committee and comments such as that the committee is *stacked* so that the result would support the change. For obvious reasons, in the case of physical fitness, the chief would not want to appoint someone who was grossly overweight and clearly out of shape physically, for example. Although the committee would want to seek out such people's feelings and suggestions for inclusion in the debate, they should not be allowed to be obstructionists or troublemakers as members of the committee.

Characteristics of Committee Members Four characteristics to look for in the guiding coalition are position power, expertise, credibility, and leadership. It is very important to pick a committee that has all of these traits; however, each individual chosen does not have to have all of these traits. This is a team effort, and it takes a diverse mix of individuals to make sure all points of view are covered.

In addition, look for members who are optimistic and are *can do* kinds of individuals. These are the type of people who, when faced with an obstacle, look for a way to solve the issue or circumvent the problem. Find people who feel good about themselves and others. They should have the courage to say what is on their minds even if it may be different or hurt somebody's feelings. Their vision should put heavy importance on quality service to the public as the number one goal. These are results-oriented people.

Again, some opponents may claim that the committee is being stacked with *yes* members. However, the purpose of this committee is to help implement the

FIGURE 3-2 Physical Fitness Training

change, and without committed supporters it would not function properly and may fail. Remember, for each major change, the chief may have to create a new committee.

The Guiding Coalition One common technique for building a trusting team is to hold some form of carefully planned off-site meeting, sometimes called a management retreat. A task group of 8–24 senior fire fighters, company officers, and chief officers go away from normal work locations for 2–5 days with the explicit objective of becoming more of a team and becoming more familiar with the problem and its potential causes and solutions. Professional facilitators are available for this process. To make this successful, a large part of the trust must come from the team believing in a common goal. When all the members of a guiding coalition deeply want to achieve the same goal, real teamwork becomes attainable.

Selecting Goal Priorities If there are several major goals in a person's life that they would like to accomplish, such as losing weight, quitting smoking, and pursuing an educational opportunity, these should only be started one at a time. This will increase the chance of success. This is also good advice for scheduling organizational change goals.

Make your own private list and prioritize the items on it. Put some items on a hold list, waiting for an opportunity to help create the urgency. Choose one item that is major, but in your mind will be easier than the rest to achieve and is clearly justified. Creating an atmosphere in the department that expects change can be helpful in lowering resistance. This is especially true in organizations that have been very effective in resisting any changes in the past. Ever hear the saying "200 years of tradition unhampered by progress"? Creating a new tradition that expects and looks forward to change will help facilitate future changes.

Step 3: Develop a Vision and Strategy

It has been reported that great leaders all have a clear vision of where they want to take their organizations in the future. By the time these people reach a top administrative position in their organization's chain-of-command, they should have the formal education and professional training, coupled with years of experience, to envision the goals that they would like to implement.

Remember, vision serves to convince members that major changes are absolutely necessary, even though they may not necessarily be in their short-term best interest. Changes almost always involve some pain. However, if chiefs clearly understand their vision, they should be able to explain the benefits and personal fulfillment that will be apparent in the future. Chiefs should also be able to describe the negative consequences that will emerge without attempting the change.

If the administrator cannot describe their vision to an audience in 5-10 minutes without losing their interest, they need to do more work in refining and gathering data for the justifications. A good analogy can really make the need for change very clear. To illustrate the need for adequate staffing on fire companies, a sports team comparison may be useful. If two basketball teams are playing each other and one team has the typical five members and the other has four or three, which team would have the advantage to win the game? Sports teams are successful because of their physical attributes and skills of their members, and these are exactly the same characteristics that a team of fire fighters needs to be effective.

In most cases, the chief will be the source of the change proposal. It may not be an original idea, but it is their vision of the idea that is first proposed and communicated to members and elected officials. The chief will have the power to set the agenda, which means selecting those items to put on the front burner. This is probably the greatest power in a leadership position.

Professional Judgment At some point in each individual's lifetime, complex decisions will have to be made. Many of us often ignore our feelings or hunches, even though these intuitive feelings are the result of a process that subconsciously combines formal education, experience, and common sense. This is where a person needs to have confidence in their feelings, especially when solid analytical evidence for corroboration is not present. This contributes heavily to what is called mature professional judgment.

Many people in the fire and emergency services business are more comfortable with making decisions based on facts and other tangible items. However, with this process, administrators sometimes forget to account for the human engineering factor: the actions of humans that could potentially impact the solution. For example, in the original installation of battery powered smoke alarms, it was never considered that occupants would deliberately disable the batteries.

Because emergency services rely on equipment that is very impressive and indispensable, fire and emergency services officials tend to focus on getting the latest, shiniest, and biggest equipment available. Citizens, elected officials, and members often relate the competency of a particular department with how many new fire trucks it has recently purchased.

A recent fire and emergency services periodical contained 54 ads for equipment and 14 for training, an almost 4:1 ratio in favor of equipment over training. This is partly the result of economics, because the equipment manufacturers have a larger amount of money to use for ads. However, it is a combination of equipment and human resources that provides the best service to the public. When one or the other is lacking, fire and emergency services are not giving their best effort to those they are committed to protecting. The chief should keep this in mind when selecting change items to support a new vision.

When a vision becomes clear, the chief may also feel uncomfortable because it may cause some pain to others. Deep down, everyone wants to be liked by co-workers. This is especially true in the typical fire and emergency services organization where new members start at the bottom and are promoted slowly to the top. Members make many friends along this road and may not want to jeopardize these friendships.

Finally, the urgency rate needs to be high enough to find the time needed to complete the transformation. This can be a time-consuming process and administrators need to keep a close eye on the progress to make sure it is on schedule. The members supporting the status quo will try to defeat the change effort at every opportunity.

Strategies One technique that is quite common, when the goal is behavior change and the chief administrator is powerful, is the authoritarian decree, which often works poorly by itself in the long run except in very simple situations. However, as part of a total strategy, in a paramilitary organization, there are ways to change behavior of some overly resistant members using the decree and appropriate follow-up enforcement. A problem in using only voluntary compliance is the likelihood that some members will not conform to the change.

Although the chief should also use the techniques that have been discussed and others that will be described shortly to gain voluntary compliance, be prepared to use disciplinary enforcement procedures. This requires setting up an accurate monitoring system to assure compliance. If a major change is being planned that will require behavioral adjustments, be prepared for resistance from some individuals.

Step 4: Communicate the Change Vision

Administrators should attempt to simplify the explanation of their vision and goals for change. If not done properly, this can really turn people off to your vision. For example, it is always a good idea to ask someone with little or no understanding of fire and emergency services activities to read anything you write. If it makes sense to this person who doesn't know the jargon or have years of experience, then it should make sense to anyone who hears the new vision. Remember that the chief will also have to explain the vision to elected and appointed officials as well as the media and public.

> Great leaders are almost always great simplifiers, who can cut through argument, debate and doubt, to offer a solution everybody can understand.... Effective leaders understand the KISS principle, or Keep It Simple, Stupid. They articulate vivid, overarching goals and values, which they use to drive daily behaviors and choices among competing alternatives. Their visions and priorities are lean and compelling, not cluttered and buzzword-laden. Their decisions are crisp and clear, not tentative and ambiguous. They convey an unwavering firmness and consistency in their actions, aligned with the picture of the future they paint. The result? Clarity of purpose, credibility of leadership, and integrity in organization. (Powell, 1996)

Using a metaphor, analogy, and example can be invaluable for clarifying the proposed change and the justification. Remember, a picture is worth a thousand words. This is where years of experience really come in handy. A very powerful method for communicating ideas is to relate the vision's concept to something that will be very familiar with the audience. For example, many people own cars. If you relate the need to purchase a new fire engine to the familiar need to buy a new car that has too many miles and years of use, you have a simple request and clear justification.

When conveying the vision of the change along with the justification to individuals and groups, you should use all methods of communication, both written and verbal. This takes a little planning and brainstorming to make a comprehensive list and use all avenues of communication. Most people do not assimilate anything the first time they experience the new idea. Using repetition and varying the message technique, alternating between different media and communications choices, will help hammer the message home.

Leadership by example is extremely important. If this is not done right, it can send a message that the chief is all talk and not serious about change. There are many chiefs of fire and emergency services organizations who mean well when they pronounce that they would like to achieve a new vision. But because many chiefs do not have a plan or an understanding of the process, they do not comprehend all the steps that must be taken for success.

There are many examples that point out how *leading by example* either helps or can doom a change effort. A newer expression states that "if you talk the talk, you must walk the walk." This is very good advice. (See the case study at the end of this chapter for a real-life example.)

Once committed to a change vision, the administrator must monitor every detail for any signs or behaviors that are contrary to the new way of doing things. Unknowingly, many people in our society practice an adaptive behavior called *denial*. Rather than face a particular problem, they are able to deny that it exists. Chiefs should search for any signs that they may be prone to this behavior.

Watch for any inconsistencies and identify them immediately. They must be either corrected or explained publicly. If their existence is denied, then they cannot be addressed. Again, the details can kill the proposed change. For a change program to work, it must have a high level of credibility with a substantial number of the members, the public, and elected officials.

The feedback loop provides information on the implementation's success and the need for any corrections. By listening to those who are implementing the new vision it will become very clear if there are problems. Listen carefully, and remember management by walking around (MBWA) from Chapter 2. Make appropriate corrections in the implementation plan as soon as problems are identified. Do not wait until you are forced by the opposition. They may use these errors to bring down the entire change project.

"The downside of two-way communication is that feedback may suggest that we are on the wrong course and that the vision needs to be reformulated. But, in the long run, swallowing our pride and reworking the vision is far more productive than heading off in the wrong direction or in a direction that others won't follow" (Kotter, 1996, p. 100). If the opposition identifies the problems and forces modifications to the change program, the status quo forces will have won a battle, and they will be harder to defeat the next time. Once a problem has been realized with the change initiative, correct it as soon as possible. Admit the mistake or misunderstanding before others force their own modifications.

Step 5: Jump Over Barriers

How are barriers created? Many times we become so accustomed to one organizational design, policies, procedures, or equipment, perhaps because they have been used for decades, that we are blind to any alternatives. This is called tradition.

For example, two fire departments, which shall remain unnamed, share a common border. One department is completely paid and the other is a combination department. When dispatching fire companies, the paid department expects that each company will respond without delay, staffed by a minimum of four fire fighters per company. This is a typical organization of a completely paid fire department.

On the other side of the border, the combination department was created by consolidating several volunteer departments into one county fire department and hiring paid fire fighters. Generally, the paid fire fighters are assigned to stations to offset periods when adequate numbers of volunteers are not available.

Many volunteer departments, when they are alerted to a call, have no guarantee of the number of members available, the time needed to respond, or even if they will respond at all, although this would be rare in most cases. This is because most do not require their volunteer members to spend on-duty time at the station or on call in the neighborhood. More recently, a growing number of volunteer organizations have been requiring on-duty staffing by their members, and they are very successful at responding quickly with minimum staffing levels.

This same practice of alerting the fire companies, with no guarantee that an adequate number of personnel would be available, was carried over into the staffing of the newly consolidated department. Many of the paid personnel were assigned to supplement the volunteers so that some companies still remained all-volunteer and other companies were staffed with one or two paid personnel and the rest of the crew was expected to be made up of volunteers. Although they saw a need for paid personnel, they did not use them to increase the reliability of their response; they simply maintained the way it had always been done.

What is completely acceptable in one department is considered unusual in another. The completely paid department responds with full crews 100% of the time. The combination department, in contrast, deals with failures to respond, late responses, and understaffed responses that are unheard of in the fully paid department. This combination department and the residents of this jurisdiction think this situation is normal. Both departments serve populations of about 700,000 residents and share a common political border.

This most important step of jumping over barriers is to identify them. As the preceding example points out, the barrier may be invisible to the organization and its administrators because it has been institutionalized into the organization's tradition. Remember, there will always be resistance to any change, even when it will improve the organization and its service to the public.

Once the barrier is identified, convincing members, elected officials, and the public that it is really an impediment to quality emergency service becomes the paramount task. If the barrier is clearly defined, in many cases, the barrier's effect on emergency service will become apparent. Of course, identifying the problem is always the hardest part of any change process and must be the first step. Just as with any personal problem in someone's life, the first step is admitting that there is a problem.

Sometimes the administrator needs to start their plan for jumping over barriers with a blank piece of paper to analyze the problem and find solutions. In other words, administrators should assume that they are creating a new organization from the ground up without any of the current, traditional barriers.

Look at Someone Else's Experiences Many local structural barriers vary from department to department. Generally, some of these barriers tend to follow what is done in a major central city; the city's procedures become the norm throughout the surrounding suburbs. However, there are some differences because of historical context. For example, when many central cities first created their paid fire departments, labor was relatively inexpensive, they were able to hire and staff their companies with large crews, staff many stations, and provide separate ladder companies with full crews in many stations.

Other than these major historical differences, good ideas can be found hiding in any fire and emergency services organization that can help overcome structural barriers. Great ideas can be found in the smallest to the largest departments, including those staffed by all volunteers, all career, and combinations of the two.

Remember, some structural barriers in one department may have a solution in another that is considered normal procedure. Another department may have adopted a practice or procedure that would help overcome a barrier. This can be very helpful in selling the new change if the proposed solution actually works in another department.

This is a good job for the guiding coalition to handle. After the vision is created, the guiding coalition refines it and locates a possible solution(s).

Troublesome Supervisors Every fire and emergency services organization has some supervisors who will probably oppose any changes. To a large degree, like all of us, they are a product of their experiences. They learned a command-and-control style early on, and because that behavior seemed to work and help them get ahead in the department, it became a deeply ingrained set of habits.

Fire and emergency services supervisors are very good at taking commands and managing emergency situations. However, these supervisors can present an enormous obstacle when major changes are pursued. Some do not like change and may see the change as a personal attack on their safe, predictable, comfortable environment. Easy solutions to this sort of problem are difficult to find.

Sometimes special interest groups, such as labor unions and volunteer associations, take on this same characteristic of resisting change. The following applies to both organizations and individuals.

If particularly influential supervisors or groups are not confronted early in a change process, they can undermine the entire effort. For powerful organizations, if the chief places representatives on the guiding coalition, which is a normal consensus-building technique, this may take care of some concerns or at least temper any resistance. If they are a part of the process, they are exposed to the justification for the change and it becomes more difficult for these organizations to strongly oppose the change. Consensus-building techniques can be very helpful in these situations.

Confront the individual supervisor or manager as early as possible in the process and attempt to convince them to buy in to the new vision. If the explanations fall on deaf ears, then notify the supervisor that any overt resistance or effort to organize defiance will be met with disciplinary action. Because of the typical merit personnel system or union agreements in government agencies, separation may not be an immediate option. However, be very clear (in writing) that any interference or noncompliance will be treated as insubordination and disciplinary action will be taken. After making this warning, disciplinary action must be carried out quickly when a violation is detected. When infractions occur, do not overlook or deny these problem members.

Step 6: Create Short-Term Wins

Almost any long-term goal should have periodic intermediate checkpoints to monitor progress. For example, when people want to lose weight by dieting and exercising, they may set their goal for the next 12 months. But as anybody who has been successful at losing weight and keeping it off will tell you, they check their weight often. With every new pound that is lost, the dieter has obtained a short-term win.

Short-term wins are absolutely necessary to carry the momentum and urgency for a long-term project. This is especially true with any *behavioral* change. Many of the real changes in organizations are behavioral. For example, if the department has adopted a goal of being more customer-oriented, this is a big behavioral change for the traditional members and must be constantly encouraged and reinforced.

A short-term win has at least these three characteristics:

1. It's visible to the entire organization and members can see for themselves whether the result is real or just hype.
2. The results or outcomes are clear and unambiguous.
3. The results or outcomes are a direct consequence of the change.

Short-Term Gimmicks Many gimmicks are the result of a misunderstanding of the proper items to measure, a failure to follow up on progress, or a deliberate attempt to mislead elected officials and the public. This is what is typically called *smoke and mirrors*, indicating something that is illusory or not real. Sometimes departments issue new SOPs and then fail to follow up on compliance. These departments will announce to elected officials, the public, and other emergency services organizations that they have a new state-of-the-art program, but in reality nothing has changed.

For example, a department has started a new program to test all fire hydrants. This program is the result of several incidents when fire companies attempted to use a fire hydrant and found that is was defective. This new program requires the detailed submission of written reports that note the location of each fire hydrant tested and any defects. The majority of company officers carried out their duties and completed the tests. However, one officer filled out the reports and did not do any tests. One very important short-term goal is to verify that everyone is participating. This is another example of where MBWA can be very effective. Chiefs cannot sit back in their offices and hope that all officers have the same understanding, motivation, and loyalty that the majority of the officers have. There will always be a small minority that will be hard to convince.

It is important is to have a plan for both short- and long-term goals. Remember, the opposition is tradition and status quo. "In enterprises that have been around for decades [350 years for the fire service], the granite walls can be thick. Sometimes, extremely thick" (Kotter, 1996, p. 130). But, with the right plan, patience, follow-up, and perseverance, the administrator can be successful.

Step 7: Be Prepared for Resistance

Tradition, as well as irrational and political resistance to change, never fully disappears. Because of this reality, be careful when celebrating or rewarding short-term wins. The opposition may acquiesce to the new change, but they are always ready to reassert their opposition. Until the changes have truly been institutionalized into the organization's culture, which may take many years, permanent change can be very fragile.

Also, look for interdependent systems that require other changes. For example, if the new goal is to inspect every commercial occupancy once each year, this may require changing other demands on members' time while on duty. Some changes may be personal, such as time spent on leisure activities, which is common in some departments.

One common interruption may be frequent responses to emergency dispatches for companies that are very busy. In some cases, these calls may be driven more by policy than real need. For example, some departments dispatch three or more companies to *automatic alarms* when usually the first arriving engine handles most of these calls. A commitment to one particular change can lead to questions, re-evaluations, and additional changes. The department may want to study the actual usage of the different companies sent on each type of alarm. Those units that have a low percentage of usage could potentially be removed from being dispatched on the initial alarm.

Whenever a new item is found that needs changing to support the original change, this insight will help solidify the first original change. This process is called *facilitating*, and the administrator should be the main person to help make possible the change using their influence. When new obstacles are discovered, the chief must help find new solutions to overcome the obstacle. Significant change takes time and patience.

Step 8: Finalize and Institutionalize the Change

The underlying culture of the organization must change if the new direction is to be permanent. The culture of an organization is very difficult to address directly because it becomes an invisible part of the normal day-to-day operations. It has a powerful influence over the behavior of members who are expected to conform. In most organizations, individuals are selected and trained to meet the expectations of the existing culture. This is done routinely by the incumbents without a conscious effort and is difficult to challenge or even discuss.

For example, over the past few years there has been a realization that although firefighting is a very dangerous occupation, many fire-fighter deaths are preventable. The NFPA created a standard for safety in 1987 (NFPA 1500, *Standard on Fire Department Occupational Safety and Health Program*), and many national leaders and organizations have attempted to influence the fire service into following this safety standard. However, in a recent study sponsored by the U.S. Fire Administration, the executive summary announced: "Although the number of fire fighter fatalities has steadily decreased over the past 20 years, the incidence of fire fighter fatalities per 100,000 incidents has actually risen over the last 5 years, with 1999 having the highest rate of fire fighter fatalities per 100,000 incidents since 1978" (USFA, 2002, p. 1). Clearly, the culture has not changed in some organizations.

What Works The following six items will help the culture change process be successful:

1. The leader and the guiding coalition need to talk a great deal about the rationalization, justification, and evidence that support the change.
2. Speak about the old culture, its founding, and how it has served the organization in the past, but is no longer efficient, safe, or useful.
3. Encourage existing members who are not open to change or do not agree with the change to be tolerant or leave the organization.
4. Lead by example.
5. Ensure that new hires are not being screened and trained using old norms and values.
6. Make sure those to be promoted are not inherently opposed to the new changes.

Cultural change must come last and is the final step in securing the change. The true test of a real leader may be what occurs after he or she leaves the organization. If the new practices and vision remain and the organization does not regress to past practices, then the full transformation has been completed.

A Final Word about Leading Change

There are ways to lead when you are not the official appointed chief administrator of the organization. Although this is not as easy as being the head person, it can be done. Delegation can go in both directions, both up and down. These are skills that can be learned from others' experience and writing. These interpersonal skills are not discussed in this text, but they are real and are great techniques to use.

Leadership is truly a lifelong learning project. The more knowledge the administrator has, the easier it will be for them to find suggestions and ideas to overcome opposition. Ask questions and reach out for the opinions of others. Also, the more information the administrator has, the better chance they will not be blindsided by a fault in their reasoning or data gathering.

> Listening with an open mind, trying new things, reflecting honestly on successes and failures—none of this requires a high IQ, an MBA degree, or a privileged background.
>
> 1. Risk taking: Willingness to push oneself out of comfort zones.
> 2. Humble self-reflection: Honest assessment of successes and failures, especially the latter.
> 3. Solicitation of opinions: Aggressive collection of information and ideas from others.
> 4. Careful listening: Propensity to listen to others.
> 5. Openness to new ideas: Willingness to view life with an open mind. (Kotter, 1996, p. 183)

Organizational change will never be easy, and you will need plenty of patience to follow the eight-step change process outlined in this chapter. But progressive change is the only avenue to creating a truly competent professional fire and emergency services organization.

Keep in mind that the process begins only after the problem has been identified. Be very careful to take the time and effort needed to assure that your analysis and its conclusions are supportable when challenged. This will be the hardest step and the most critical.

CASE STUDY: Leading by Example

A metro fire department adopted a physical fitness program that required mandatory participation and a yearly test. This program covered new employees only, and they were all required to sign a *condition of employment* document that they would not be able to stay employed as a fire fighter if they did not pass the test. Because these fire fighters had to pass this test to graduate from their recruit school, they were all capable of passing the test the first day in a fire station.

Once each year, all battalion chiefs had to oversee a physical fitness test for the fire fighters in their battalions. One particular BC was a strong believer in keeping fit for both personal and professional reasons and also was able to successfully pass the yearly test.

During each shift the BC would participate in the department's physical fitness program and on many occasions would join one or more companies during their training. It was well known by all fire fighters how the BC felt about physical fitness. For the yearly 1.5-mile run test, the BC would advise the fire fighters that they would have to beat him to the finish line. The BC would then run at a pace to finish in 12 minutes, which was the pass-fail mark. In the 3 years that the BC was assigned this field command, *he never won a race*. In this period of time, all the fire fighters passed the test.

Discussion Questions

1. Is this battalion chief helping to lead change? If yes, explain in detail.
2. Is it fair that the new employees have to pass this test and not the existing fire fighters? Discuss fairness in terms of the new recruits, fellow employees, the department, and the public.
3. Find (using an Internet search) and analyze three separate ideas or arguments that would support a physical fitness program for fire fighters (or would support a no exercise viewpoint). Provide complete reference sources.

References

Kotter, John P. *Leading Change*. Boston: Harvard Business School Press, 1996.

NFPA. *National Fire Codes*. Quincy, MA: NFPA, 2003.

Powell, Colin. *Quotations from General Colin Powell: A Leadership Primer, 18 Lessons from a Very Successful American Leader*. http://www.usna.edu/JBHO/sea_stories/quotations_from_general_colin_powell.htm (accessed May 17, 2003), 1996.

USFA. *Firefighter Fatality Retrospective Study*. http://www.usfa.fema.gov/downloads/pdf/publications/fa-220.pdf (accessed July 22, 2005), 2002.

4

Financial Management

Knowledge Objectives

- Comprehend the budget process and planning.
- Understand budget types and terminology.
- Be able to analyze government budget documents.
- Examine tax revenues, financing, and alternative funding sources.
- Comprehend strategies to counteract adverse financial impacts.
- Examine changes in public sources for funding government budgets.

What Is a Budget?

The budget document links resources to expenditures, translates financial resources into services, provides accountability, and can be a measure of the efficiency and effectiveness of the money spent. This chapter will explain the tools that can be used to identify the real impacts of increases or decreases in a budget along with some techniques to help achieve adequate funding for the department's goals and objectives.

There are many types of budgets, and each has it strengths and weaknesses. It is up to the administrator to understand the budget process. In most cases, department officials, the public, and elected officials will not be able to measure success in achieving goals in the fire and emergency services department by looking at reports that are designed for accountants, not administrators. Budgets are about money, not goals or outcomes.

Although chiefs will never be experts on taxes, they must have a good understanding of all tax issues before proposing a tax solution to help fund a new program or enhance an existing one. Taxes and their financial impact will be discussed at some length in this chapter.

Line Item Budgets

There are several types of budgets. One of the easiest to understand and analyze is the line item. This type of budget is very good for controlling expenditures and tracking financial accounts; however, it is not very good at measuring real results or outcomes.

Line item budgets are generally reported on a monthly basis, and it becomes obvious when expenditures are either too high or low compared with the allocated funds. For example, one line item that always needs monitoring closely is overtime. If after 6 months more than half of these funds have been spent, a closer review of these expenditures will be needed. A thorough check of the justification and appropriateness of the expended overtime would be necessary along with projections of these expenses for the remainder of the year. If the overtime was justified and unavoidable, then funds will have to be acquired to cover the shortfall. In this example, the chief could request a supplemental appropriation, transfer funds from other line items, or cut back on overtime for the rest of the fiscal year.

The line item budget can make it simple to justify a yearly increase, because an administrator can easily add an incremental amount of money to cover increases in costs. This is true in good financial times, but the opposite can also be true. If the tax revenues decrease for the next fiscal year, it is easy for the government accountants to simply cut all budgets by a specific percentage. This relieves them from cutting a specific program and thus they do not have to take the heat for cutting a popular public service. What chief would like to close a fire station or reduce staffing as part of a budget cut?

To illustrate the complexity of a reduction in funding, consider the following example. The accountants in the finance department or elected officials have proposed to cut the budget, and the chief is given the latitude to propose a reduction plan. If the department had 100 employees and now the new authorization would only be 90, in implementing this budget cut, the chief may choose (hypothetically) to cut all of the fire prevention and training staff positions, which total 10 employees. This proposal would keep all existing fire stations open. In many real-world cases, this is the least controversial plan.

In the short term, the elimination of fire prevention and training may not have a big impact. However, as the months and years go by after the cut, the real impact will become evident as fire fighters lose their proficiency, and injury rates may increase. Also, there will be greater numbers of and more severe fires. In many actual cases, the politically expedient cuts in budgets follow this example.

Methods of Increasing Revenues

"Show me the money!" This was the theme in the movie *Jerry Maguire*, in which a football player was looking for the big payoff from his next contract. It is not uncommon for sports players to have salaries larger than many fire departments' entire budget. Governments, unfortunately, cannot negotiate a big increase in their revenues overnight. A good understanding of revenues (primarily taxes and fees) is needed to be able to protect the budget and be successful in obtaining increases for new programs.

There are basically two ways to receive a substantial increase in the fire and emergency service budget. Because all revenues end up in the same big pot of money, the chief can always hope that their department's new program is so popular or justifiable that government officials will transfer funds from other agencies to fund the project. Generally, it is not a good idea to go after somebody else's budget to fund a new project for your department; however, if an opportunity presents itself where the elected officials have become convinced that a program in another agency is no longer necessary or justified, be prepared to step in with a proposal on how the money can be better spent.

Another funding source for new programs can come from an increase in a tax or fee. These increases can be generated through the approval of elected officials or increases in revenues based on the normal fluctuations of these sources. In many cases, elected officials are not big fans of raising taxes; therefore, the chief must convince them that there is an overwhelming need for the new program. The goal is to present a solid justification that would be supported by the voters. The elected officials normally will not go against the wishes of the people who elected them, especially at the local level. If this connection cannot be made, the funding request will probably fail.

Property Taxes

Property or real estate taxes are the most common funding sources for local government programs, particularly in states that do not have a state income tax. This tax is assessed as a percentage of the total value of property and improvements such as a house, or in some cases a rate based on income for commercial property.

The property tax is relatively easy to collect. Nobody can hide a house or a Home Depot store from the tax collector.

For a property that has structures on it, there is a direct connection with fire services because there may be a need to extinguish a fire in the building(s). This is not as clear with other public services such as education, police protection, and emergency medical services, which are based on the occupants, not the building.

However, because this is a flat rate as opposed to a progressive tax like the federal income tax (more on flat vs. progressive later in this chapter), all property owners pay the same percentage of the value of their property. If a homeowner lives in a million-dollar house on a horse farm, she or he will pay a lot more tax than a family living in a mobile home on a small parcel of land. The greater the value, the greater the tax.

Commercial properties may have different rates and can be based on the income of the business as opposed to the actual value of the property. Fluctuations in the property tax can be the cause of decreases in tax revenues. In parts of the country, when property values fell, so did the tax revenues. This reduction usually does not occur quickly because the tax only changes after the property is reassessed by the tax collector's office. Some states assess properties on a multi-year basis; therefore, the impact will not be realized for several years. There is a lag between the cause and the effect. This is also true on the other end of this trend, when the property values start increasing. This type of shortfall in tax revenues can be predicted ahead of time by monitoring the sales price of properties in the community. However, in many cases the elected officials will be hesitant to act until an actual financial crisis has occurred.

Income Taxes

Another type of revenue source is the income tax, which is the federal government's primary source of funds. This is a progressive tax that is characterized by some as being fairer than a flat tax because it taxes at a higher percentage the taxpayers who make higher incomes. The federal tax laws allow many deductions that are predominantly used by people with higher incomes and have the effect of reducing the total amount and percentage of tax paid.

However, in 2000 the Internal Revenue Service (IRS) reported that the top half of U.S. taxpayers pay 96% of all federal income tax. Included in the top 50% are taxpayers who have an adjusted gross income of $27,682 or greater, which probably includes many paid fire fighters. Because of this apparent inequity, many higher income taxpayers have pressured Congress and the president for tax cuts and have been successful.

Several states link their state income tax to the federal calculations for taxable income. Some states do not allow many deductions and have a straight percentage rate. Some cities and counties have an income tax that is based on a percentage of the state income tax. For example, a taxpayer may have to pay an additional 50% of the state income tax to the county.

Some cities also have income taxes. For example, a fire fighter in Detroit, Michigan, must reside in the city, which has a 3% income tax.

The income tax has one major growing problem: tax evasion. Tax evasion amounts to approximately 22%–23% of all income taxes collected, and this does not include taxes lost from illegal sources of income, mainly cash in the underground economy. This is an ever-increasing problem and has been reported to be growing at over 6% per year (GAO, 1994).

The income tax can have its ups and downs. If the economy goes into a downturn or recession, this source of funding will be affected very quickly. All government agencies will be quickly thrown into a crisis mode.

Sales Taxes

The sales tax will suffer the greatest immediate negative impact during poor economic times. One of the first things people do when faced with the loss of a job or the threat of being laid off is to stop spending on items that bring in the sales tax, such as cars and large appliances.

The sales tax is a common source of revenue at both the state and local level. Most states have a sales tax. Some states, such as Alaska, Delaware, Montana, New Hampshire, and Oregon, have no sales tax. For the other states, there are differences in both the percentage rate and the items covered from state to state. For example, some items that may be exempt are food, clothing, and medicine. In several states, either a county or city government can add a local sales tax on top of the state's sales tax.

Some states like the sales tax, especially those that have a lot of tourists. Tourists can pay a substantial amount of tax to help finance state and local governments. The main services these people use are roads, police, and fire. Tourists do not send their kids to the local schools, which is the major cost for state and local governments, so state and local governments aggressively pursue tourism. There are many examples of state or local governments reducing or eliminating real-estate taxes on a new amusement park or tourist attraction in anticipation of the sales tax receipts and new jobs.

Other Taxes and Fees

Another tax that is not very common but is used in some areas is the fire service tax. This tax can be one value for all dwellings or can vary depending on the square footage of the dwelling. It can also be calculated by using a *fire flow* formula and applied to commercial structures as well. This formula is commonly used by fire officers to determine the amount of water in gallons per minute needed to extinguish a fire and varies with the combustibility and area of the structure.

One tax that is very popular is the use tax or fee. One of the most common use taxes is the road tax, which is added to the price of fuels used by motor vehicles. Consumers see this as a fair tax because it is directly based on the use of the roads. The more miles driven and/or the poorer the fuel economy of the vehicle, the more tax that is paid.

These taxes are usually directed back to the cost of the service. Technically the government does not make any profit from these taxes over the cost of the service, but it is common for there to be more tax brought in than is needed. For example, a portion of the federal road tax is being diverted to offset some of the national debt.

To help pay for public services, many jurisdictions charge a use tax on motel rooms, rental cars, and admission tickets for amusement parks and tourist attractions. In theory, this is to offset the cost of public services provided to tourists in these sites.

Governments use many different types of fees to offset the cost of specific services when those services benefit the user directly. For example, road tolls, driver's license fees, professional licenses, and ambulance transport fees are very common. Some fire departments become sold on taking over EMS transport after calculating the amount of revenues that can be realized from transport fees.

Some fire prevention agencies charge fees for plan review and inspections. The department should be careful about being the government's enforcement authority (fire inspections) and charging a fee for this service. For example, some legislation requires periodic inspections of an occupancy or fire protection equipment installations, and then requires that a fee be charged for the inspection. These fees should not be used as a way to acquire additional funding for the agency's overall needs, but just to cover the actual costs of the inspections.

Borrowing

Funds for major capital improvements are generally provided by loans through the sale of bonds. A capital item is typically an item that has a long life expectancy, such as a fire station. Fire stations have a 30- to 50-year life expectancy and are very expensive, so they are a perfect example of a capital purchase item that must use funds that are borrowed from the sale of bonds.

Government bonds will have interest rates that are lower than the rates for commercial money, but the purchaser of these bonds will receive their interest tax-free. These bonds are therefore a good deal for both the investor and the government.

Many governments have either an official or a recommended debt ceiling. Many times this is expressed as a percentage of the total revenues, such as the bond debt payments should not be more than 10% of the tax revenues. Several governments have suffered financial crises due to the inappropriate use of bonds.

There are two types of bonds that can be used by governments, short and long term. Long-term bonds are those that have a redemption time frame of 20 to 30 years. Short-term bonds can be as short as overnight to several months.

To bridge the gap in the receipt of tax revenues, some municipalities have used short-term borrowing. For example, July 1 may be the deadline for the payment of the local real estate tax. As a result of fiscal mismanagement, on June 1 the government will run out of cash. Therefore, the city will temporarily borrow funds to cover its expenses for that month. This can end up being a troublesome practice, as the 1-month loan extends to 2 months, then 3, and so on. This was one of the questionable financial practices that brought New York City to the brink of bankruptcy in 1975. More recently, in November 2002, the New York City Council passed a property tax increase of 18.5%—the largest in the city's history—to help close a multi-billion-dollar budget shortfall.

In these cases, the government is borrowing to cover operational costs such as salaries. This is a poor practice. The options for the elected officials are to either cut costs or raise taxes. These are two actions that most elected officials dread.

The federal government is able to borrow to cover its excess expenditures over revenues. This is called the federal debt. Theoretically, this debt never has to be repaid; only the interest payments must be paid each year. Therefore, the federal government and its elected officials can vote to spend money, but do not need to vote for taxes to pay for the spending.

Smaller fire and emergency services departments have used bond sales to raise funds to purchase fire rescue apparatus. With the cost of fire and rescue apparatus at levels of $250,000 to $1 million, it is very difficult for small departments to be able to save that amount of money to pay cash. Because these departments will be buying only one new fire rescue apparatus every 5 to 20 years, a bond may be a good method to raise the needed capital.

Larger departments generally have purchase plans to buy a certain number of apparatus each year and should stay away from borrowing or leasing. Any time money is borrowed there will be a cost. (Depending on actual interest rates, bond money will cost between 15% and 20% more over the term of the loan.) For example, if the department has 10 engines and 3 ladder trucks, the most economical purchase plan over the long run would be to budget and purchase one fire rescue apparatus each year. With this plan the department will have a number of pieces of equipment that are relatively new and others that are near retirement.

A few larger departments have used a bond sale and replaced all (or a large portion) of their units at one time. In these situations, in year one the department has all brand new equipment, but as time goes on, all of the apparatus will age simultaneously, increasing the probability of repairs, breakdowns, and out-of-service time.

With all the apparatus aging at the same time, downtime for repairs will become excessive and a crisis will materialize. There are those who argue that the political system works best in a crisis mode. They believe that the political process resists spending money until an obvious crisis exists, especially when budgets are tight. These are the times when a politician can take credit for solving a problem.

A replacement plan that budgets to replace apparatus on a scheduled basis is the easiest to justify. Once approval for this type of funding is gained, it will be easier to achieve approval for the next fiscal year. If last year's budget contained $250,000 for a new engine, then next year's budget request would just include an incremental increase to cover the increased cost for one year. It is far easier to gain a 3% increase in the budget than to go back and justify each new apparatus purchase.

Lease-purchase is another type of financing that can be used to acquire expensive assets such as fire rescue apparatus. This type of financing does not affect the debt ceiling; if the government has a high percentage of debt, then this may be the only option. Again, generally this will cost more than a cash purchase when calculated over the term of the lease; however, if there is a real crisis need for a new fire rescue apparatus, this may be the only option. The lease-purchase may have a better chance of approval from the legislature and the financial officials because the costs become a line item in the operating budget instead of a capital purchase.

In many jurisdictions, the authority to issue bonds must be approved by voter ballot. This can take a long time and there is no guarantee that the voters will approve the bond issue. However, fire bond referendums historically have the best chance of approval and will receive the highest percentage of yes votes from the voters when other bond issues are on the same ballot.

Investing funds is another source of revenue for governments. Governments have the ability to earn interest on their bank accounts. Many sources of revenue for governments, such as taxes, are due to be paid by a certain date. For example, the federal income tax is due by April 15th each year. Most of this revenue will be needed to pay salaries and other fixed costs throughout the year. Because these funds will not be needed immediately, the government can invest in short-term instruments. Many larger governments hire financial managers to invest these funds.

Some investments can be very risky, especially those that report high returns. Generally, governments tend to stay with low-risk investments. For example, the federal Social Security Trust Fund can invest only in federal government bonds (T-bills). Even in times when some investment funds were reporting up to 30% returns, the Social Security Trust Fund was making only 5% or 6% on its surplus funds.

However, one notorious exception to safe investment by government was practiced by Orange County, California. On December 6, 1994, Orange County became the largest municipality in U.S. history to declare bankruptcy after its risky investments lost most of their value unexpectedly.

Caution is always necessary with government revenues. Many states are cutting their budgets and dipping into emergency funds to make ends meet. In some situations, this is being compounded by the reduction of investment income in retirement accounts. When states should be contributing more into retirement accounts to offset the losses in stock investments, they have less tax revenues.

Basic Economic Theory and the Relationship to Government Finance

The gross national product (GNP) is the total output of the nation's economy including wages and salaries of employees; the profits realized by entrepreneurs, industries, and stockholders; and the rents received by landlords. Over 30% of the GNP is created from the spending of local, state, and federal governments. These are huge sums of money.

However, almost 70% of government spending is done by the federal government, and it is rare that any of the federal government's spending would be for local fire and emergency services. Most of this federal spending is for entitlement programs, such as Social Security, welfare, and Medicare.

One interesting concept is that of *tax expenditures*. At the local level, this can be seen in the practice of exempting certain properties and their buildings from the local real estate taxes. For example, state and federal property along with nonprofit organizations are typically granted these exceptions.

Fire and emergency services are a public service that all property owners, visitors, and residents have unrestricted access to because local governments or nonprofit organizations typically provide them to anyone who requests them. Therefore, what really happens is that the property owners who do pay real estate taxes essentially subsidize free public fire and emergency services for those properties that are tax-exempt.

Public emergency services can charge fees to offset this inequity, such as an ambulance transport fee. In some jurisdictions where state or federal governments or nonprofit organizations have large holdings of property and buildings, a concern with fire officials is that these organizations do not pay anything toward the public fire protection they receive.

Other problems also are created because there is typically no charge when the citizen calls the fire department. This public service will be requested more times than it is actually needed simply because it is free. For example, many fire departments are struggling with the problem of nuisance and false automatic alarms. As long as the service is free, there's no incentive to correct any of the problems that cause these false alarms. To control this problem, some jurisdictions have adopted an ordinance that fines businesses that have multiple false alarms. This encourages the business to fix the malfunctioning alarm system.

Many departments that provide EMS transport are struggling with heavy demands for service. This service is essentially free to the patient in most cases. Even where there is a fee, it is either paid by medical insurance or Medicaid or not paid at all. A free public service can create a market inefficiency when people request the public good (emergency ambulance service) when it is not really needed.

Another example is apparent in multi-family dwellings without smoke alarms. Both the individual purchasing a smoke alarm for their own apartment and their neighbors will benefit from the smoke alarm. If a fire starts in an apartment without a smoke alarm and nobody is home, it will not be noticed until the fire has grown substantially and has a good chance of extending to adjoining apartments. The people in the adjacent apartments would want a smoke alarm in the apartment where the fire started, but they have no direct influence over that person's actions.

In some cases, there would be a disincentive for the owner to maintain the detector because of numerous false alarms. Therefore, whatever public good a smoke alarm could do for all of the apartment building's occupants, some individuals may not see it as being in their best interest to buy and install one. In this case, market forces cause a level of smoke alarm use that is less than the efficient quantity (i.e., less than all apartments), causing a dangerous situation for everyone: more fire deaths, injuries, and property damage. This is a good example that would justify government regulation.

Another example of externalities occurs when an individual building owner considers the installation of an automatic sprinkler system. Hopefully, the owner will consider the benefit of lower insurance rates and the survival of the business if he or she were forced to shut down because of fire damage. However, in many cases the developer will opt to forgo the extra expense of a fire sprinkler system. If a sprinkler system is installed in a building, this would produce a public good, eliminating a potentially large catastrophic fire in the community and saving the lives and property of anyone in the building.

A fire of large magnitude may overwhelm the local fire department's resources and present a hazardous situation for fire fighters. Also, if it destroys the building, the municipality will lose a substantial amount of real estate tax, not to mention the number of local people who would be without jobs if the building is a business. One of the major justifications for automatic sprinkler ordinances is the elimination of target hazards that would have this type of adverse economic impact on the local community. These are benefits to the entire jurisdiction in addition to the property or business owner. Again, when market forces fail to bring a demand for public good, this becomes a justification for government regulation. This subject will be discussed further in Chapter 9.

Budget Process and Planning

There are typically several distinct steps that constitute a budget cycle for fire and emergency services organizations.

Planning: This is done at the agency level in most cases. A formal planning process can be a great help. In the formal process a task force is selected with representatives of the agency, the public, other agency heads, and elected and appointed officials. Documents such as National Fire Protection Association (NFPA) 1500 and 1710, Occupational Safety and Health Administration (OSHA) regulations, and the Insurance Service Office (ISO) Fire Suppression Rating Schedule are excellent guides for the task group. A through discussion of planning is found in Chapter 11.

Budget preparation: Requests are prepared on standard forms provided by the budget or finance agency. Justifications must be prepared for each major section of the budget. Any new items or enhancements will call for detailed justification and possibly their own separate budget form. For guidance, the chief should acquire the budget requests from previous fiscal years and other agencies in the government. Review very closely any successful requests for substantial increases in budgets that another agency submitted in previous years. In many cases, the administrator will receive some type of guidelines from either the city manager, county manager, or budget director that may detail items such as the exact percentage increase or decrease expected or proposed for next year's budget. This is especially true with salaries and any cost of living increments.

Budget justification and adoption: In most cases, the proposed budget will receive a preliminary confidential review by the budget director and city manager. This process could vary depending on state and local laws. Some areas with Sunshine laws will require that this preliminary proposed budget and justifications be made public. However, there will always be some informal private input from those officials. When the city manager, public safety director, or budget director speaks about items they would like or not like in the budget, the chief should listen very carefully. It is important to resolve any differences before making budget wishes known publicly.

All agency budgets will be consolidated into one municipal budget. The budgets will be analyzed for accuracy and compliance with municipal or agency policies and objectives. If the total requests are above the estimated revenues, then it is typical for the personnel in the budget office to make recommendations on budget cuts for individual agencies. The chief needs to stay close to this process and use every opportunity to express support for the fire and emergency services budget.

Do not forget the art of negotiation and compromise in dealing with the personnel in the budget office. It is always a good idea to have a trustworthy relationship with the officials in the budget agency. These relationships are not built up overnight, but are generally the result of long-term friendships.

After the budget receives the blessing of the municipal executive officer and the budget agency, it is then officially transmitted to the legislative body. Hearings are scheduled on the budget and it is made available to the public. At this time, the chief and department staff members should be prepared for inquiries from special interest groups, the media, and citizens groups as well as the labor organization. Remember, at this point the budget is now a proposal of the senior elected or appointed official, and any attacks on this budget need to be defended.

At some point in time, the chief will be scheduled to appear before the legislative body and officially present the department's budget for adoption. The chief should be very familiar with the budget and its justifications. Some chiefs will go through a role-playing exercise with their senior staff. The staff will be instructed to go through the budget and ask the most difficult questions that can be dreamed up. The chief will also want to have senior staff at the budget presentation and defer to staff members when the question is in their area of expertise.

Budget administration: This is the final item of the budget cycle. Expenses and revenues are monitored in the budget and are revised, if needed. Periodic reports are prepared as required by the executive officer, elected official, or budget department head.

The Role of the Fire and Emergency Services Financial Manager

The financial manager's role in a fire and emergency services organization is a busy one. When one budget cycle is finished and the budget has been adopted and funded, the next cycle starts. The manager must immediately monitor any potential adverse impact on the budget such as unexpected overtime, increased costs in capital items, revenues below predictions, and unfunded mandates.

If the budget was approved on the basis of a fee collection, keep a very close eye on this collection process to make sure it does not fall below expectations. A revenue shortfall in fee collection will more than likely directly impact the agency. A revenue shortfall in a municipal tax can be remedied, however, by a reduction in any or all of the municipal agencies' budgets, not just one.

Financial managers should review any proposed legislation and regulations from state and federal governments. Look for any potential fiscal impact on the fire and emergency services organization. Set up an automatic referral system that will acquire copies of proposed legislation and regulations. This area of planning can be made a lot easier by networking with peer financial managers and other fire and emergency services organizations. Also, several professional organizations, such as the International Association of Fire Chiefs, periodically publish summaries of proposed legislation and regulations, especially those proposed by the federal government.

In smaller organizations, the financial manager may actually be the chief administrative officer (chief) of the fire and emergency services organization. In larger organizations, the chief will delegate this monitoring of proposed legislation and regulations to a staff officer. Clearly instruct staff to report in writing the existence of any proposed legislation or regulations that may have a financial impact on the department.

Look for opportunities throughout the year to communicate with citizens groups, the press, and elected officials to gain support for existing programs or enhancement programs that may be proposed in the future. When these opportunities occur, they're very short-lived. Be prepared with all the information for justification and an accurate estimate of all costs.

Monitor the community for demographic changes that may affect the services requested and tax revenues available. For example, older communities may see both a decrease in population and a decrease in the average income per taxpayer. This is double trouble. Suburban areas experiencing growth will realize increases in tax revenues. At the same time, however, there will be strong support for increased spending in areas such as education. Most growth areas will see an appreciable number of new families with children.

Some areas are seeing increases in elderly populations that do not have any children of public school age. This type of population growth can be advantageous because the new tax revenues will not have to be used for education.

Voting and Public Choice

How do governments theoretically make decisions about budgets? Keep in mind that public choice is not as systematic as private choice, such as when a person purchases a new automobile. Governments must choose between many conflicting interests. It is the job of the fire and emergency services chief to put up the best arguments to support the department's budget requests. If successful, the chief will have convinced the elected and appointed government officials to spend taxpayers' money on the department's needs.

Governments do not always do what is right, efficient, and in the best interests of the citizens or the government. In the area of fire and emergency services, it is the chief's duty to attempt to convince the government's elected and appointed officials to provide the organization with the funding needed to provide the best fire and emergency services.

Direct voting is a process whereby voters are allowed to vote on government decisions as opposed to the more common process where voters elect representatives. For example, direct voting often is used in many small New England towns to approve budgets, ordinances, and bond issues. This is true democracy. For the majority of towns, cities, counties, and states, elected representatives vote for the citizens who elected them to manage the budget and other government matters.

It is assumed that when all else is equal, voters will vote according to their pocketbooks. One example is the difference between voters with and without school-age children. Those voters with school-age children would be more likely to vote for a bond issue for a new school building. When bond authorizing referendums have a number of different public services requesting approval, it is common for the fire service request to receive the highest percentage of yes votes. This can be the result of the public's trust and that all voters may have a need for fire and emergency services.

Again, the public generally thinks the fire service is doing a good job, so it may be difficult to convince them that there is a need for additional resources. However, by making a convincing argument, the public generally will trust that the requests are justified.

So far this discussion has been in terms of voters, assuming that most elected representatives will behave in a like fashion. This is particularly true at the local level. Elected officials are concerned with re-election. If they watch the pocketbooks of the voters very closely, they will have a good chance of being re-elected.

When requesting additional funding, voters or their representatives will have to be convinced that spending the extra funds will be of some value to them. The justifications should contain a very clear explanation and evidence that the department will be able to increase the quality of fire and emergency services provided with the additional funding.

For example, if asking for funds for a new fire and emergency services station along with the staffing for a new fire company and paramedic unit, the chief will have to demonstrate in the presentation and written justification that the increased spending will provide better service for the citizens in the new service area. In this case, the better service is easily measured by comparing the existing response time to the projected response time when the new station is constructed.

Be prepared when an opportunity becomes available by having a list of construction projects, increases in personnel, and other needs for the organization. A master plan is an excellent source for these needs and their justifications.

Other opportunities to gain support may surface from other nearby fire and emergency services organizations. For example, a department may need a new training center and would also like to hire additional training instructors. For smaller organizations, the cost of these enhancements to their budget would not be approved because of the relatively high costs. But, if the costs could be spread among several fire and emergency services organizations, then the smaller request would be more likely to be approved by elected officials.

This same technique, partnerships, can be used for other services that are very expensive and infrequently used. A regional approach (on a county or several counties basis) to services such as hazardous materials, terrorism planning, response to local catastrophic events, and technical rescue teams is a win-win situation for a cooperative effort among several fire and emergency services organizations. Each fire and emergency services organization's contribution in equipment and personnel becomes more reasonable when shared. In addition, the state or federal government may be able to help fund these regional teams through Homeland Security programs.

Bureaucrats and Other Agency Managers: The Competition

The administration of a fire and emergency services organization is part of the government bureaucracy. Bureaucrats from other agencies are also in competition for tax revenues. These bureaucrats are expected to implement the programs approved by elected officials. "But implementing a political program is not like executing a computer program. The bureaucracy is not a simple machine that blindly follows orders" (Bruce, 1998, p. 205). Therefore, bureaucrats end up with a lot of latitude to determine how to implement their programs.

If a budget cut is proposed for the education department, the agency head may propose to eliminate a school sports activity, for example, even though the school superintendent may know there are some unneeded staff positions that could be eliminated without any adverse impact on the district's prime mission—teaching students. Instinctively, the school bureaucrat knows that there will be opposition from parents (voters) and students to the sports activity cut. This will give them the opportunity to have the funds restored during the budget approval process. Where will the elected officials find the extra money to transfer to the school budget? Raising taxes will probably be a last option.

As previously mentioned, however, raising taxes does occur, as it did in New York City when the real estate tax rate was raised by over 18% in 2002. The additional revenues reduced the proposed cuts to the Fire Department of New York for 2003, but the department will still have to close six fire companies. This may not save its budget in the long run, however, because spending for schools in New York City is predicted to skyrocket in the future.

Only relatively recently (in the 1970s and 1980s) have standards become available that can be used to benchmark services provided by fire and emergency services organizations (and, therefore, help justify budgets). In the past, most fire and emergency services organizations set their own level of services provided to the public in a kind of ad hoc fashion, without any formal planning.

Most government agencies have very few standards for the services provided to the public. For most, the output of government departments is very difficult, if not impossible, to measure. However, government administrators have better information about the operation of their departments and their cost structures than do elected officials. The elected officials are, to some extent, at the bureaucrats' mercy.

If the police chief proposes a budget increase to cover the cost of new police officers, the chief typically would justify this by reporting that the department would be able to reduce some types of crime. The police chief may select a crime that had recently received some notoriety. Because there is no standard on the number of police officers needed, the chief's request will probably be judged by the elected officials on the potential for gaining votes in the next election. Remember,

> ... the bureau manager associates personal and professional success with success in obtaining the largest possible appropriations or grant for his or her department ... because government does not maximize profits, bureau managers do not have an incentive to reduce the cost of production and operation ... and because many government outputs are difficult to measure, bureau managers' performance is assessed on whether they have followed all the rules and correct procedures, not on whether they have delivered good value to the taxpayers. (Bruce, 1998, p. 206)

This may also explain the previous administrator's behavior in the budget process. Today, there are some relatively new national standards that were not previously available. These standards should help focus the budget request on quality service to the public. In general, the three areas that are covered quite comprehensively in the national standards are:

1. *Training for fire fighting and the emergency medical services personnel:* NFPA professional qualifications standards and Department of Transportation standards for EMTs and paramedics
2. *Safety of personnel:* NFPA 1500, OSHA Fire Brigade and Respiratory Protection standard
3. *Deployment and staffing of fire and EMS units:* NFPA 1710 and ISO Fire Protection Rating Schedule

Therefore, the more chiefs rely on national consensus standards, a formal planning process, and/or cost-benefit analysis, the better chance there is to convince the elected officials to support new and existing programs.

Cost-Benefit Analysis

In most cases it is very difficult to measure benefits or direct outcomes from fire and emergency services programs; however, there are some benefit evaluation techniques that may be helpful.

Can a dollar value be assigned to a person's life or quality of life? If a survey was performed about the value of fire and emergency services, there probably would be a great variance in the valuations. For example, a family enjoying a nice quiet evening in front of their TV, with no fire or smoke in their home, would probably express a low value for having a fire and emergency services organization ready to respond and rescue them in a fire emergency. Take this same family and place them into a 13th-floor apartment with a fire blocking their exit to the stairway, and there would be a new higher value for the fire and emergency services organization.

What makes cost-benefit analysis even more difficult is the realization that the outcome, saving lives and property, is the result in most cases of a complex set of circumstances. The following is a hypothetical set of circumstances that led to a death in a house fire:

1. Fire prevention regulations did not require smoke alarms to be installed in existing dwellings.
2. The local fire and emergency services organization did not offer any home fire safety inspection services, or the free installation of smoke alarms.
3. The occupants of the dwelling did not see any value in purchasing and installing smoke alarms.
4. The occupants of the dwelling did not recognize the hazard of careless smoking habits in the family room.
5. The first arriving fire unit was understaffed and/or personnel were not adequately trained.
6. When the victim was eventually located and brought to the front yard, the fire and emergency services organization was not equipped to provide emergency medical services.

It could be argued that correcting any one of these six items could have potentially saved this fire death victim. This makes it almost impossible to construct a valid cost-benefit analysis.

Although some of these items are clearly more expensive than others to correct, programs can be implemented to correct all the deficiencies indicated. Anything less than 100% effort in all areas will leave open opportunities for life loss and extra property loss as a result of structural fires.

Funding

One good technique that chiefs can use is to tell staff not to bring any problems to their office unless they also bring some suggested solutions. The problem may be a need for additional funds for fire rescue apparatus, stations, personnel, incentives for volunteers, and any other budget items. When requesting enhancements, the additional funding can be transferred from other agencies' budgets or, with some luck, there may be a big increase in tax revenues projected for next year.

Be prepared to recommend a funding source from a new tax or an increase in an existing tax or fee. Proposing a tax increase should be saved for those occasions when attempting to make substantial improvements in the organization and its services. This may also be used to gain funds to implement part of a master plan. For example, in a department that needs three new stations for adequate coverage, an implementation plan may detail the opening of one new fire/rescue station a year for the next 3 years. This will also require the hiring of new personnel each year. The proposal for funding may include step increases in a tax to generate an additional number of dollars each year.

Along with a proposed funding source that is a tax or fee, be prepared to explain all the ramifications and who is affected. In general, if the taxpayers are convinced that there will be added value or service from the increase in the tax or fee, they will probably support the increase.

Examples of fees that have been approved in the past that are directly related to the services provided are ambulance transport fees and impact fees for new development. In the case of the transport fee, the value of the service is very clear to most people. The impact fee is a little more complicated. One justification can be that existing fire stations and other equipment costs have already been paid for from previous tax revenues. The new development will have the advantage of these buildings and equipment, or may need new facilities to protect their property. More commonly, public buildings and infrastructure (typically, schools, roads, and fire stations) need to be constructed to serve the new development along with sewer, water, electricity, telephone, and cable TV.

Taxes

"Taxes are what we pay for civilized society," said Supreme Court Justice Oliver Wendell Holmes Jr. in 1887. What is a tax? What types of taxes are there? When is a tax fair? And what impact will the tax have on the local economy?

There are two primary reasons for government taxes: to bring in revenue and to influence behavior. For example, the government may use a high excise tax on cigarettes to reduce consumption. The government has a financial interest in reducing cigarette smoking, because older cigarette smokers need substantial amounts of expensive medical care. Cigarettes are the direct and indirect cause of a large number of deaths and serious illnesses in the United States. This increases the cost of programs like Medicare, Medicaid, and private medical insurance. Elected officials do realize, however, that when smokers quit, they then will live many years longer and have a healthier life. This will have increased financial impact on the Social Security system and other retirement programs that have to pay these people more as they live longer lives. Many public policy issues have multiple facets to their problems and solutions, and many are eventually solved with some type of compromise.

Because all taxes are levied on some market activity like income, wealth, or sales, all taxes have an economic impact. This means that many businesses, special-interest groups, and the taxpaying public follow very closely any proposed changes. To illustrate the magnitude of tax revenue that local, state, and federal governments collect every year, each person in the United States would have to pay an $8,500 lump sum payment if they had to pay an equal share.

Taxes have an economic impact on the market; let's look at the sale of a bagel as an example. When either an excise or sales tax forces the consumers' price for bagels to go up, the demand for bagels will decrease. This is a classic example of supply and demand—higher prices decrease the quantity sold.

Also, a tax on the manufacturer or seller of bagels will have the same effect as the tax paid by the consumer in the case of an excise or sales tax. The ultimate effect is that the cost per bagel will rise and the sale of bagels will fall.

Every product and service has a price-demand curve. FIGURE 4-1 demonstrates a classic curve representing the supply and demand relationship for bagels.

It can be very difficult to estimate tax revenues. For example, let's say that the city has a 10% sales tax on bagels. Last year, the local bagel store sold 100,000 bagels that resulted in tax revenues of $10,000. To fund a new program, it is estimated that an additional $10,000 will be needed. Therefore, an increase in the bagel tax to 20% is proposed, in the hopes of doubling the tax revenues to $20,000.

After the adoption of this increase in the sales tax, the first year revenues are below the predicted total of

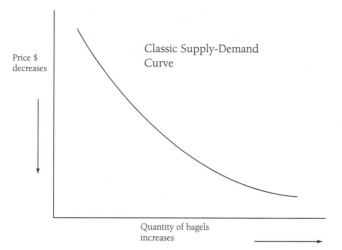

Price $ decreases

Classic Supply-Demand Curve

Quantity of bagels increases

FIGURE 4-1 Classic Supply-Demand Curve

$20,000. What happened? The supply and demand curve in **FIGURE 4-1** shows graphically the results of increasing a tax and answers this question. When the price of bagels rose because of the additional tax, the demand for bagels fell and a reduced number of bagels were sold. This resulted in a loss of some of the previous tax revenues that were generated at the higher sales numbers.

In fact, there is a tax rate above which tax revenues will actually decrease. Also, in some cases, when a tax is lowered, tax revenues can go up. This is a very complicated subject and requires a lot of study to master.

In addition, the effect at the local level of new regulations may be the loss of potential tax revenues. For example, if taxes are raised on new housing, or the cost of new housing increases as a result of new regulations in the local jurisdiction, then some would argue that the potential customers might purchase their homes in other localities. This is only partially true because in many cases the overwhelming reasons for purchasing a particular home are the characteristics of the neighborhood and the schools. Therefore, unless the new total cost is substantially increased, the increase in taxes or costs as a result of regulations will have a weak impact on home purchases.

The other important point is the explanation about who actually pays the tax, the consumer or the producer/business. A good example occurred in 1996 when Congress reinstated the 10% federal excise tax on airline tickets. Even though the price of airline tickets should have gone up 10% to cover the new tax, the airline tickets rose only 4%. In other words, the businesses typically have to absorb some of the tax levied on their products and services. In this case, the net price received by the airline companies fell by 6% as a result of the airline excise tax. This same re-

sult may not hold true over a longer period of time, and eventually the prices to the consumer do increase.

In this case, as well as for all taxing issues, businesses do not actually pay the tax. For the airline industry, the 6% loss in revenue will be passed on either to the investors in reduced dividends or to employees through reduced salaries. If you follow the money to its source, you will always find an individual who actually pays the tax. For example, when buying a car from a large motor company, any federal corporate taxes will be included in the cost of the car (approximately 22%); this will reduce the dividends to stockholders and limit employees' salaries.

There are other ways governments can raise the price of a commodity that are not taxes, but act like taxes. At the local level, when there are new requirements in a building or fire safety code, most of these costs are passed on to the consumer. For example, if there is a proposal for a new regulation requiring fire sprinkler systems in all dwellings, the housing industry will argue that raising the cost of new homes, particularly for the first-time buyer, will keep these buyers from buying because they cannot afford the increased price.

Tax Incidence Analysis: Who Bears The Tax Burden?

It seems obvious that the person who actually pays the tax is the one who incurs the adverse economic impact of the tax. However, in many cases this is not true, and in some cases may be only partially true. For example, the owner of a multi-family rental building pays real estate taxes on this property each year. But does the owner really pay, or are the renters indirectly paying the tax in their monthly rents?

It is very important to be able to understand who actually pays. If the administrator is proposing a new tax or an increase to an existing tax or fee, many of the arguments in opposition to the tax or fee will evolve around who actually pays. When making a proposal to raise the real estate tax on rental apartments, be aware of how much the landlord will pay of the tax and how much the renters will pay. If the renters are predominantly low income or elderly, the landlords and the renters will argue that the renters will pay all the new tax. In the public forum, this would give them the best chance of defeating the proposal.

When a tax is being judged for its equity, the ability to pay is a common measure of fairness. Those taxpayers who are wealthy and have higher incomes, it is argued, should pay a greater share. If the tax does this it is called *progressive*; if it doesn't, it is called a *regressive* tax. The federal income tax is a progressive tax.

When considering whether to recommend that a progressive tax be adopted, keep in mind that the people in the highest 10% of the income bracket probably have a lot of influence in the local political process, and they pay over 67% of the federal income tax already. In many cases these people would rather have no new taxes because the arguments to offset the fairness issue are not very strong.

The real estate tax is viewed by some as a slightly regressive tax. This tax is calculated at a certain rate per thousand dollars of the value of the property and any improvements. Even though the rate does not change or increase with the value of property, as would be the case with a progressive tax, it is typical for the wealthy to own the more expensive properties and buildings. In addition, another popular tax, the sales tax, generates more revenue from the wealthy because they buy more and their purchases tend to be more expensive.

In some cases, an increase in a tax or a new tax will change taxpayers' buying and spending habits because they are trying to avoid the tax. For example, if the government levies a tax on fuel oil for heating, in the short run, home and business owners with oil furnaces will not find it valuable to replace their serviceable oil furnaces with more efficient units or furnaces that use different energy sources. The consumption of fuel oil is reduced only as a result of turning down their thermostats to lower temperatures or adding new insulation to buildings. However, in the long run, the oil-burning furnaces will grow old and deteriorate, and many will be replaced with other heating options because of the higher priced fuel oil.

This may be a good outcome. Remember, taxes can be used to influence behavior. With this new oil tax, the consumption of fuel oil will be reduced, and because most oil is imported into the United States, this tax will reduce dependency on foreign oil. In the long run this tax will succeed in its goal, which is to change buying and spending habits.

Fairness and the Tax System

If the increase or the new tax has the appearance of not taking into consideration the taxpayer's ability to pay, then the proposed tax will more than likely have an uphill battle for approval.

If an increase in the real estate tax is proposed, there may be a lot of opposition from those people on fixed incomes. For every dollar collected in new taxes, the low or fixed income taxpayer will have to reduce spending an equivalent amount on essentials such as food, shelter, and medicine. In many tax proposals the impact is a complicated issue. Wealthy people who own very expensive homes and properties will end up paying the majority of any increase in a real estate tax; however, those people who are on low or fixed incomes will also be hurt by these increases. Some jurisdictions have added items to the taxing ordinances to reduce or limit the burden on the truly needy.

Another test for fairness examines if the benefits or public good are truly worth the additional cost. If the benefits derived create an increase in service or an improvement in the quality of the service, then there may be strong support from taxpayers. A good strategy would dictate that a tax increase should result in noticeable improvements in service. There are a large numbers of taxpayers in this country who feel there's not any good justification for an increase in taxes. They distrust government at all levels. However, remember that the fire service is trusted more than any other public institution and may be able to overcome the public's fears.

If there is a chance of overcoming these negative government feelings, it is with a strong justification of improving service. There must be a clear cause and effect relationship between the new program's spending and increases in service or the quality of the service. Arguments that are ambiguous or subjective will often fail.

There are cases, however, where common sense arguments do seem to work. For example, if requesting funds for a new fire engine, the simple, common-sense argument that a new fire engine is more reliable than the existing older fire engine often will not be refuted. But be prepared to add statistics to support the common-sense argument, such as any experience with downtime for repairs

Tax Avoidance and Evasion

A desirable feature for a tax system is low cost of administration and compliance. For every tax there are taxpayers who try to minimize or circumvent tax payments through avoidance and evasion.

There are numerous opportunities to avoid paying federal income tax. Some are legal, and some are not. A clever taxpayer can find many loopholes that will bring taxable income down substantially. These methods can be very complicated and require an intimate understanding of the U.S. government's tax regulations and may require the use of accountants, financial planners, and attorneys. In general, the federal government experiences a substantial amount of tax loss due to a variety of tax evasion techniques. Some are very complicated and some are as simple as not reporting income paid in cash.

An example of tax avoidance at the state or local level is evident when consumers purchase items over the Internet or through catalogs based in other states to illegally avoid the sales tax in their home state. Consumers have also avoided local sales taxes by driving to neighboring states with a lower or no sales tax to make purchases. The sales tax added at the retail level is a lot easier to enforce than income taxes, however.

Again, the one tax that is almost impossible to avoid is a real estate tax. The property and buildings are fixed and can't be moved or hidden from the tax official. In most cases, it is also very easy to find the owner. Thus, as long as the assessment on the property was done correctly and legally, this is one of the easiest taxes to collect and the hardest to avoid.

State and local property taxes are classified as taxes on wealth. Typically, property taxes are calculated using a levy rate or mill rate per $1,000 of valuation on the property and its improvements. It is common in many jurisdictions to have a separate rate for commercial buildings that is determined by the income production of the business or a different mill rate.

Rising values and property tax rates have resulted in some tax revolts across the country. The first and most famous was California's Proposition 13 in 1978. This proposition limited the growth of the property tax rate and set a maximum limit for the levy rate. It also created a situation where the assessment is only changed, beyond the limited growth allowed in the proposition, when the property is sold, as compared to most states that reassess properties on a scheduled basis without any limits. Over the many years since implementation, this has caused many inequities. For example, long-time owners pay less tax simply because they have owned their homes longer. Because homeowners who have been in the same house for a number of years have a lower real estate tax than more recent purchasers, many homeowners have been hesitant to sell their homes even when other reasons would normally encourage them to move, such as a long-distance commute to a new job.

The property tax rate also affects developers' decisions as to where to construct a new building. This can be of great concern to the appointed and elected officials because in most cases they would prefer to attract new development and its new taxes to the municipality. Although the property tax rate is not the only criterion, it does have some influence over the economic viability of the project. Some jurisdictions reduce, suspend, or eliminate the property tax to attract a major industry.

Sometimes other economic factors are of greater concern, such as the availability of public water and sewer. For example, a proposal for a new shopping mall may end up within a municipality's borders simply because the city will provide public water and sewer (a city-owned and -operated utility) only inside the city. For a single-family homeowner, this will not be as important because a well and septic system can be installed at a relatively low cost.

There are situations when the higher property tax may seem worthwhile. For example, parents with school-age children will support higher taxes in a city with an outstanding public school system. There are many who would move to a city for the benefit of better schools or lower crime rates. It is very unlikely that people move for any other public services such as fire or EMS. Homes are purchased in neighborhoods that have some type of status even in areas of high property taxes. In addition, retirees and senior citizens are often attracted to areas with medical facilities.

Buildings burn and the fire department will be called to extinguish the fire, so there seems to be a direct correlation between the fire service and the property tax. However, the value of the property does not necessarily relate to the costs of fire services needed. For example, a valuable industrial complex pays a large property tax bill every year, but because it is protected by automatic sprinklers, the probability of needing public fire services is slight.

In some areas of the country that have fire protection districts (5,601 total), a dedicated property tax is used exclusively for fire and emergency services. Normally, property taxes are collected by the municipality and spent on many different services. This puts fire and emergency services into direct competition with agencies such as police and education. It also guarantees that the request to spend public tax dollars must be justified and be of a higher value than other government services. If chiefs are good at the budget game, they will get their fair share if not a little more.

The fire protection districts have some problems even though there is no competition for their share of the tax. Some are set up with an elected board of fire commissioners who manage the tax revenues. In some areas of the country, the fire commissioners who are elected are also volunteer fire fighters from the organization for which they will now oversee the spending of the fire tax proceeds. Although most are well-meaning, public-spirited citizens, it is often hard for them to be as objective in their oversight of these public funds as would be elected officials who are not fire fighters. Although many volunteer organizations (as well as many career and combination departments) are not funded properly, some of the fire tax districts are spending the public's money on unnecessary items without proper scrutiny. These situations are analogous to a hypothetical city turning over the management of the fire department's budget to the fire fighter's union. The union officials would make the determination of how these funds would be spent. What would happen if this group of union members had to make a decision on whether to purchase a new fire engine or to fund a salary increase for fire fighters? It is not that these people are bad or corrupt; they have simply been placed in a difficult situation.

Throughout this text, strategies will be identified for acquiring a fair share of the tax dollar with the understanding that there needs to be some checks and balances when using public funds. Remember, *absolute power always corrupts absolutely.*

Fairness

If a property tax rate increase is proposed, one of the arguments the opposition typically uses revolves around the issue of *fairness*. Do these taxes impose an undue hardship on those with low or fixed incomes? It would be easy to point out that the wealthy and commercial property owners end up paying the greatest percentage of the property taxes, but that is only because they have the most expensive properties and homes.

One method for correcting this inequity is to provide a homestead credit, as is done in Florida. Each homeowner is allowed a $25,000 exemption from the assessed value of their property and home. This has the effect of exempting many homeowners from any property tax because Florida has many RVs, mobile homes, and manufactured homes that are valued at $25,000 or less. Therefore, these homeowners do not pay any property tax in Florida even though they may use public education, police protection, and fire/rescue services. One solution to this inequity has been the use of a fixed-fee tax per dwelling unit for fire and emergency services. Because fire and emergency services are very closely connected to calls for service from the occupants of the structures, and some dwelling units pay no property tax, it seems logical and fair to have these people pay a per-dwelling unit fee. Property owners and elected officials are more comfortable with this type of tax because it is very clear that it goes for a service that directly helps the owners and occupants.

In general, state and local governments are limited both legally and practically in their ability to increase these taxes on consumption (sales tax) and wealth (property tax). States cannot legally tax anything involved in interstate commerce. They also cannot discriminate against out-of-state people or businesses in their application of taxes.

Businesses and homeowners can choose where they want to reside. If the taxing situation in one state becomes excessive, they can move to another state, so there is some competition to keep taxes low to attract and keep businesses and new residents. There are states that aggressively lower or temporarily suspend taxes to recruit new industry. There are also some states that do not seem to be concerned with the amount of taxation. Several states are now finding it very difficult to balance their budgets and may have reached a point where tax increases are not a solution to their budget shortfalls.

A situation that keeps some taxes low can be seen in areas with large tourist industries. Taxes such as hotel, entertainment, and rental car taxes are typically paid by nonresidents.

Property taxes are the major source of revenue (75.6%) for local governments and their fire and emergency services departments. Therefore, if additional funds are needed to support existing programs or start a new program, the property tax and user fees will probably be the primary sources. Taxes that require minimum enforcement effort, such as the property tax, are those that work best at the local and state level.

Remember, increases in taxes or fees can be successfully achieved when a proposed new program is chosen that will create an increase in quality or quantity of service and the tax or fee is fair, equitable, and easy to enforce.

Economic Impacts of Taxes on the Fire and Emergency Services Organization

Most fire and emergency services organizations are funded at the city or county level. The tax revenues at the local level can be impacted not just by the local markets, but also by state, national, and even global markets and economic trends.

The taxes that typically support fire and emergency services organizations are the sales and property taxes. Anything that hurts the local economy will impact the collection of these taxes. The exact timing of any economic impact on the local budget can vary depending on the local tax collection process.

In 1970, America imported 6.1% of the GNP and exported 6.8% of the GNP. By 1993, these numbers had more than doubled to 14.3% for exports and 17.2% for imports. More and more of the U.S. economy is affected by global influences and financial trends. Problems with the stock markets in Indonesia, Japan, Russia, Brazil, and other countries have had substantial impacts on the U.S. economy. Recently, terrorist attacks by organizations and individuals from other countries have had a substantial negative economic impact.

Trade agreements will open up U.S. markets to competition with even more foreign products. If there is a local industry that provides a substantial contribution to the tax base, there may be some adverse economic impact if this industry has new competition from foreign countries. Or, the local industry may be able to increase their production to supply a growing demand in the global market.

The two trade accords that the U.S. has entered into are the General Agreement on Tariffs and Trade (GATT) and the North American Free Trade Agreement (NAFTA). Both have the same effect of opening up trade between the United States and North American/European countries by either eliminating or lowering tariffs on imports. Previously, a country could apply a high tariff (essentially a tax) on imports to make foreign products more expensive, thus giving the local products an advantage by having a lower price.

This, combined with the general shrinking of the American share of the world market, creates a situation that can impact the national economy, which can have an effect at the local level. America's balance of trade shifted in 1976 to a deficit situation that has grown into billions of dollars annually. This is the balance sheet that indicates that the United States buys more foreign products than it sells abroad. Therefore, the flow of money from the United States to other countries is an ever-increasing drain on the nation's wealth.

The United States is the world's largest debtor nation. Although this situation does not seem to have any real impact on the economy at the present time, keep this potential adverse economic situation in mind.

A 1999 news story carried by Reuters stated that economists believe "The biggest problem facing the U.S. economy is the risk that inflated stock prices could suddenly tumble and trigger financial market volatility . . ." (Reuters News, 1999). This prediction came true in 2000 and has caused a substantial reduction in local and state tax revenues that may force changes in fire and emergency services agencies' budgets. In addition, the tragic terrorist attacks of September 11, 2001, and the corrupt executives of several major companies added to the financial troubles of the United States during this same period of time. More recently, the U.S. economy has been doing really well and tax receipts have rebounded.

The United States has lost the lead in the manufacturing of technology consumer goods. For example, there are no domestic manufacturers of televisions or DVDs. The majority of low-end semiconductor chips are manufactured overseas. America is becoming a service-oriented economy.

If the local community relies on industry or manufacturing for a good portion of its tax base, there may be trouble for these companies in the future and trouble for taxes. The decline in manufacturing jobs is a natural evolution of the labor market in this country. As the costs for labor have risen in the United States, the costs of labor for manufacturing in other countries have stayed at a lower level. It is cheaper to manufacture a product overseas in many cases.

This same evolutionary process took place in the United States when machines took the place of manual labor in agriculture. For example, where many laborers would be needed to process cotton, the cotton gin (invented in the late 1700s) was able to process 50 times as much raw material. Many common laborers were replaced with the new machine.

The American educational system is receiving low marks for its ability to produce graduates that excel in high technology. The U.S. economy is now entering the technology service–oriented age and schools are not able to produce the skilled, educated workers needed to compete in the global market.

For example, Sumitomo, a Japanese company located in Durham, North Carolina, hired high school graduates to serve as quality control technicians, just as this company does in its Japanese facilities. The new employees could not master the statistical analysis required in the job, so the job's educational requirements were upgraded to a college degree. Newly hired college graduates also could not perform the job satisfactorily. The company finally had to hire graduate students and statistics majors to meet the job standards for a job performed by Japanese high school graduates.

Be aware that major changes are always possible in the economy at the local, state, and national level and may happen without notice. Be prepared to protect the budget from an assault by the municipal administration and other department heads when and if a downturn in the economy occurs.

The U.S. economy is highly regionalized. For example, an older city that contains a large number of heavy manufacturing firms may experience a greater reduction in local taxes than the suburban area that surrounds the city. Older cities in the United States have already seen reductions in city services as a result of declining tax revenues.

A similar trend can be seen in U.S. industry as companies downsize and re-engineer themselves. Many of the jobs lost are in the middle management ranks. These people are taxpayers in somebody's municipality. The key management description for the newly changed organization is leaner and meaner.

In general, the public is showing more and more resistance to increases in taxes to pay for improvements or expansion of public services. In some cases taxes have been reduced by voter referendums. However, ". . . the total U.S. tax burden is equal to 56 percent of annual personal consumption spending" (IPI, 2001). This includes federal income tax, Social Security and Medicare tax, and state and local taxes such as income, property, and sales taxes. Also included are hidden taxes such as 22% of the cost of a new car or truck. And each year it continues to grow.

Remember that some of the most costly federal programs are not fully funded. Social Security and Medicare are projected to run out of money in the future. Some of the options to maintain them are to either raise taxes or reduce benefits. Because most federal elected representatives are not anxious to take either option, these problems may not be resolved until a crisis occurs, and then change will be unavoidable.

In many fire and emergency services agencies costs are increasingly driven by reductions in work hours and increases in salaries and benefits. At the same time, increases in staffing levels are being requested, which requires more employees. Many of these increases are justified, but these trends must be considered when protecting an existing budget, and even more so when requesting enhancements.

Many fire and emergency services organizations have produced improvements in productivity by offering new and expanded services such as EMS and hazardous materials mitigation. In some cases these improvements in productivity have caused increases in response time to structural fires as fire and emergency services units find themselves with heavy workloads as a result of EMS calls, which have grown to 61% of all emergency calls (NFPA, 2004).

For volunteer fire and emergency services organizations, economic pressures have reduced the availability of members due to increases in overtime and second jobs. This is especially true for older members who find themselves with new demands on their time as they start families. With the increase in dual-income families, both spouses have fewer uncommitted hours that could be available for training or to respond to alarms.

Changing social values have influenced many men to take a more active part in the raising of children. In past years, men were free to commit many hours of their time to the local volunteer fire company (**FIGURE 4-2**). Between the increases in demand caused by the expanding roles of the modern fire and emergency services organization and the new values of a stronger commitment to the family, men are no longer staying active after taking on the responsibility of a family. Some of this downward trend has been offset by an increasing number of women volunteering, however.

For additional insight, the U.S. Fire Administration (USFA) report *Recruitment and Retention in the Volunteer Fire Service* discusses the core problems of recruitment and retention in the volunteer fire service, and provides suggestions for solutions to these problems from volunteer fire chiefs and fire fighters from across the United States. This free publication can be ordered from the USFA Web site (http://www.usfa.fema.gov).

Alternative Funding Sources in the Fire and Emergency Services

Fire and emergency services organizations can take advantage of some creative sources of revenue if there is a justifiable need for increases in quality or expansion of emergency services.

The USFA report, *Guide to Funding Alternatives for Fire & EMS Departments*, provides information on locating and implementing both traditional and nontraditional methods of funding. It discusses local, state, and federal government sources for financial help.

A recently implemented federal grant program offers a new level of financial assistance to fire departments in the United States. In 2003, FEMA and the

FIGURE 4-2 Volunteer Fire Fighters

USFA awarded more than $495 million to over 6,600 fire departments throughout the United States under the Assistance to Fire Fighters Grant Program. This grant program was started in 2001 and was scheduled to be eliminated before the September 11, 2001, terrorist attacks occurred. Initially a stand-alone program, it is now being folded into the U.S. Department of Homeland Security's consolidated grant program management.

The USFA has a Web page that helps departments with their grant requests. Insight into how to be a successful grant applicant can be obtained by contacting other departments that have been successful. In addition, the evaluation of the grants is done by fire service peers. Find out who are the most successful applicants and ask them for advice on a grant application.

Purchasing and Procurement Processes

Purchasing policies vary widely among different governments and their agencies. A dollar limit for a small purchase may be anywhere from $20 to $1,000. The signature of the agency head may be sufficient to authorize these small payments.

Generally there are many and varied polices and procedures for purchasing. The chief must become familiar with local rules and procedures in this area. Failure to follow these rules has been cited in numerous disciplinary procedures. This is one area of financial management in which it does not pay to be too creative.

If there is not a formal written policy for a particular purchase, find out what the past practice has been and document the procedure. Document actions by sending a note to the budget analyst assigned to the department. A good test is to ask if the transaction would look unusual to an outside observer such as a local news reporter.

The following are those areas the administrator should be familiar with:
- Different dollar limits depending on the types of commodities
- Standard procurement of expendable items
- A dollar limit that requires a formal bid process
- Contract items on a master purchase order list
- Petty cash policy
- Signature policy for purchases
- Property or equipment inventory procedures
- Capital items authorization and approval policy
- Travel policy and approval procedures

- Any standard service contracts
- Any lists of prior approved vendors

Purchases can be justified by a *useful life span* prediction for the item. For example, if the department has four fire engines, and the chief can convince the governing body that the *useful life* of a fire engine is 12 years, then a good policy would dictate purchasing a new fire engine every 3 years. This would assure that no engine would be over 12 years old.

Another calculation that should be familiar to the administrator is total cost purchasing. In these calculations, the following costs are determined, calculated, and totaled:
- Cost of the item
- Maintenance costs during the life of the item
- Downtime costs that may include the rental, standby, or reserve equipment requirements
- Surplus value for those items that have a value at the end of their useful life
- Warehouse or storage expenses
- Delivery costs

When selecting a bidder for a major purchase, using this calculation may help you to justify choosing a vender who is not the lowest bidder. If the low bid vendor's equipment has high maintenance costs or worse than average down time, the total costs may be more than another bidder with a higher purchase price. Remember that the business of emergency services is about always being ready to respond. An unreliable piece of equipment should not be acceptable. Contact peers and get a list of others who have purchased the same equipment and check out their experience with the manufacturer.

Look for opportunities for the joint purchase of equipment or supplies with other local or state governments. In some areas, items such as hose, protective clothing, and even fire apparatus are on a regional purchasing bid list. These joint arrangements consolidate purchasing power and result in lower costs. For example, the department may be able to take advantage of a state bid on staff vehicles, a county bid on fire engines, and an adjoining city bid on fire hose.

Riding someone else's bid can save a lot of money. Typically, any item that is bought in quantity will have a better price. A small fire and emergency services department may have to pay a much higher price than a big city organization. The majority of fire service organizations in this country can be characterized as being small departments. Out of the more than 30,542 fire organizations in the United States, only 110 departments protect populations over 250,000 (NFPA, 2004).

However, there are some possible drawbacks to cooperative purchasing, such as a compromise with local preferences, a potentially longer time for delivery, and a lack of control over the contract process. Most of these disadvantages can be overcome by detailed review of the specifications, noting where they differ from the department's previous purchases. Then, if the variance is truly significant, a request for a modification to the supplier or manufacturer may be obtainable. Also, become involved in the process for the development of the consolidated bid specifications; the department will be happier with the results.

There are many legal considerations to the purchasing process, and the chief needs to understand the rules, policies, and laws for the local jurisdiction. In some jurisdictions, an affirmative action plan to encourage the selection of minority-owned businesses must be followed for evaluating bids. Some of these processes have been changing recently, however, due to newer legal findings.

The U.S. Supreme Court has found *set-asides* unconstitutional in some cases. This is a process that requires a certain percentage of the purchasing dollars to be awarded to minority businesses. Even though these practices may not be acceptable in a court of law, many jurisdictions still use these techniques for awarding contracts and bids for political reasons. If the administration and elected officials in the city strongly support these practices, it would not be a good idea to attempt to change these practices except on a case-by-case basis. If a bidder who is not awarded the contract files a civil lawsuit against the city, the city may lose the lawsuit or settle out of court.

Unfortunately, some of these practices will cause the spending of additional dollars from the budget. In private conversations, point this out to officials and see if these conversations can influence the bid award process or acquire additional funding for the budget. When discussing these issues, argue from the perspective that the department needs the best equipment or supplies that can be purchased within reason. Stress that emergency services to the public require the most reliable equipment and supplies possible. Avoid discussing the legal aspects or ethical considerations.

Some chiefs have worked for many years with a contractor, supplier, or manufacturer that they prefer. Take care not to allow passion or loyalty for a particular brand, model, or contractor to interfere with the established rules for bidding and awarding of contracts. Many bidding regulations and policies do allow some subjective judgment when two bids are relatively close.

For example, in the past the department has only purchased ABC fire trucks, but now XYZ manufacturer has presented a more appealing bid in response to the latest request for proposals. The department may prefer ABC because the department has standardized on this equipment; they need to store only one manufacturer's parts and the department's mechanics have become very familiar with maintaining and repairing this brand, which simplifies the maintenance program. Using total cost calculations may show equal or reduced costs over the life of the equipment.

Budget Reconciliation

Fire and emergency services managers should be able to spot mistakes or unplanned expenses as soon as possible so that spending can be adjusted. A system should be set up that keeps track of expenditures on a monthly basis.

At the beginning of the fiscal year, a spreadsheet (either electronic or paper based) will need to be assembled that shows the expected expenditures on a monthly basis (**TABLE 4-1**). Some expenses, such as salaries, are very predictable. Others may vary by the month, or be a one-time expenditure. Each line item will be broken down into expected monthly costs.

After each month, the expenditure for that month will be noted along with any variance from the expected. Also, a running total will be calculated by line item along with any variance. It is extremely important to not spend more than the allocation in the budget. Would an employee like to have their paycheck bounce because of insufficient funds? This would also not be good for a chief's job tenure.

When committing to a purchase or service, budgeted funds can be encumbered. This is an action that reserves monies until the invoice is received and payment is issued. For example, your department may contract for the testing of ladders. At the beginning of the year, the funds would be encumbered so this budget item will not look like it is still available, even though the testing and the subsequent bill may not occur for several months. In a budget shortfall, any line item that looks like it will have a surplus will be evaluated for a transfer, and, yes, it may be to another department.

Keeping real-time spending records for each month will facilitate the planning of any future and actual seasonal peaks (e.g., winter heating oil) or one-time costs occurring in a later month. For example, the purchasing of protective clothing for wildland fire-

Table 4-1 Fire and Emergency Services Department Budget Expenditure Report

Object	Description	Budget Plan	Current Month	Year-to-Date	Encumbrances	Balance Available	%
102	Salaries, Permanent	$695,271.00	$57,939.25	$230,975.90	$0.00	$464,295.10	33.22%
103	Salaries, Overtime	$48,668.97	$4,055.75	$13,294.27	$0.00	$35,374.70	27.32%
104	Health/Welfare Insurance	$75,228.32	$6,269.03	$24,999.00	$0.00	$50,229.32	33.23%
105	Dental Insurance	$11,402.44	$950.20	$3,907.65	$0.00	$7,494.79	34.27%
106	Retirement	$108,462.28	$9,038.52	$36,795.00	$0.00	$71,667.28	33.92%
107	Vision Care	$2,781.08	$231.76	$939.27	$0.00	$1,841.81	33.77%
120	Medicare	$2,363.92	$196.99	$786.46	$0.00	$1,577.46	33.27%
Total Category 1		$944,178.02	$78,681.50	$311,697.55	$0.00	$632,480.47	33.01%

Object	Description	Budget Plan	Current Month	Year-to-Date	Encumbrances	Balance Available	%
201	General Expense	$3,500.00	$605.00	$421.00	$0.00	$3,079.00	12.03%
241	Printing	$1,300.00	$137.50	$550.00	$0.00	$750.00	42.31%
251	Communication	$8,000.00	$608.00	$2,505.00	$0.00	$5,495.00	31.31%
261	Postage	$2,400.00	$177.00	$792.00	$0.00	$1,608.00	33.00%
291	Travel	$500.00	$0.00	$0.00	$225.00	$275.00	45.00%
331	Training	$800.00	$0.00	$0.00	$275.00	$525.00	34.38%
341	Facilities Operations	$4,500.00	$0.00	$996.00	$0.00	$3,504.00	22.13%
361	Electricity	$16,000.00	$2,310.00	$9,240.00	$0.00	$6,760.00	57.75%
371	Heating Oil	$18,000.00	$591.00	$2,110100	$0.00	$15,890.00	11.72%
410	Professional & Special Services	$45,000.00	$3,712.53	$4,418.27	$12,000.00	$28,681.73	36.49%
431	Data Processing	$1,800.00	$148.50	$594.00	$0.00	$1,206.00	33.00%
503	Uniforms	$500.00	$0.00	$0.00	$500.00	$0.00	100.00%
524	Vehicle Operations	$49,000.00	$3,791.16	$15,348.66	$0.00	$33,651.34	31.32%
569	Equipment Rental	$12,000.00	$0.00	$3,960.00	$0.00	$8,040.00	33.00%
571	Unallocated	$0.00	$0.00	$0.00	$0.00	$0.00	0.00%
Total Category 2		$163,300.00	$12,080.69	$40,934.93	$13,000.00	$109,365.07	33.03%
Total Org/Prog		$1,107.478.02	$90,762.19	$352,632.48	$13,000.00	$741,845.54	33.01%

fighting is planned in the eighth month of the fiscal year. This may be 2 months before the next wildfire season starts. Because these funds cannot officially be encumbered until the order is actually placed, the administrator would want to keep note of this future purchase in their own personal spreadsheet.

It is also a good idea to keep personal records of expenditures to double-check any computer reports from the budget department. It is always possible that there can be data entry mistakes. When dealing with number codes to designate the department and line item category, a mistake in hitting the wrong number key can put expenditures into the wrong agency's budget. A procedure of double-keying, where two different operators enter the same data, helps to eliminate most of these errors.

Check any computer reports for timeliness and accuracy. For example, the department has just re-ceived an order of fire hose costing several thousand dollars, but it has not shown up on the latest budget summary report. In the administrator's records, make sure that there was not a reported balance available in the equipment purchase line item that is not really available. At some point in the fiscal year, the budget department will catch up with its paperwork and calculate a minus situation that has to be made up from other parts of the budget. How would the chief like to explain to the union president that annual leave is being canceled for the rest of the year because there is not enough money for overtime to cover vacant shifts?

It is a good idea to keep last year's total expenditures for each month to identify abnormalities that occurred last year. These can help explain any variances noted in the present budget cycle.

Are there any unexpected costs, for example, as a result of federal or state mandates, labor contract changes (especially salaries and benefits), damage caused by severe weather, accidents, or a large-scale emergency? Look for any potential surpluses to cover these costs. Can funds be transferred between line items by the administrator, or must the budget department or the legislature approve the transfers?

At the end of the fiscal year, it is typical for many government agencies to take steps to spend any funds that would be lost at the end of the year. A common technique that is used by some budget department personnel is to temporarily suspend capital purchases towards the end of the fiscal year. Plan ahead.

One additional area in which to look for transfers is in the encumbrances. As mentioned earlier, an encumbrance is an accounting procedure that indicates that the money has already been spent even though the actual payment may be made sometime later in the fiscal year. In some cases, the encumbrance can be paid in the next year's budget, allowing the use of those funds this year. For example, funds to add compressed air foam (CAF) systems for several engines have been encumbered earlier in this fiscal year. The contractor may agree to extend the contract at the same price, allowing delivery and invoicing to occur in the next fiscal year. This is a one-time fix, and contingency funds should be provided in the next year's budget to cover the unexpected costs plus the funds for the CAF systems.

TABLE 4-1 is an example of a typical line item budget summary sheet. At this time, this department has finished one-third, or 4 months, of the fiscal year. Therefore, the benchmark for its expenditures is 33.33%. On the bottom line, the report points out that 33.01% of their budget has been spent or encumbered, which should indicate a healthy budget position.

However, there are several line items that are over 33.33%, such as printing, electricity, and travel. The travel may not be a concern if most of the department's travel had been anticipated in the first 4 months. However, in checking the department's spreadsheet that breaks down the anticipated expenditures, a substantial amount of travel is projected for the upcoming months. In this case, the administrator would want to closely monitor and study this situation. This may call for canceling travel or transferring funds from other line items.

The *Total* line depends on the individual line items above contributing to the overall expenditures. Noticing that heating oil is at only 11.72%, this indicates to the budget person that the heating season will be in the last 8 months and that any surplus in this line

item is only temporary. This temporary surplus is responsible for the on-target position of the budget.

Accountability and the Audit Process

Essentially there are two types of audits: financial and performance. Financial audits are fairly common and are done to assure that the financial records are accurate and complete. They are usually done on a yearly basis. Assets are checked to assure they are still located in their assigned place and are not missing. The approved budget is checked for compliance with actual expenditures.

Performance audits come in several different types, such as compliance, procedural, internal, and external. These audits attempt to measure and scrutinize some specific area of the operations and budget of the agency.

A compliance audit may check adherence to a legal or policy mandate. For example, the department may charge a fee for plan review in the fire prevention bureau. The audit would check on collection of this fee from the calculation of the fee to the receipt and storage of the funds. If the department has a petty cash account for small purchases, this is another common area for the compliance audit.

A procedural, internal, or external audit can cover almost any subject in the budget. These types of audits may attempt to measure management policies and procedures. Many consulting firms specialize in these types of audits.

If the department is selected for one of these performance audits, try to work very closely with the person or consulting firm who has been selected to perform the audit. The better the relationship with these auditors, the better chance for fairness and input into the final report. Remember that the reputation of the administration and the department may be the focus of the audit.

The chief would be wise to create a management team that can work with the outside consultants. Select members of the department who are familiar with the budget and represent all different functions and services that the department provides.

No department is perfect and all will have room for improvement. This type of audit can be an opportunity to make improvements in the department and its service to the public, so do not fight all negative comments and suggestions from the consultant. Look for ways to turn these suggestions into new requests for additional funding. This may be a once in a lifetime opportunity.

If the administrator becomes defensive and tries to justify all existing policies and procedures, the consultants and those that hired them (for example, the mayor or the city council) will view you as a person who would resist any changes. There are occasions when these types of audits can be used to help justify increases in the budget and this can be a good result.

When the chief finds strong resistance to budget requests from elected or appointed officials in the municipality, he or she could try to convince them that an outside consultant would provide an independent professional opinion of the budget requests. For example, the chief (and the union) may want to hire additional fire fighters to bring staffing up to safe levels of four fire fighters per company. Make sure the request is realistic before calling in an outside consultant to review the plans and ideas; for example, it would not be realistic to expect a consultant to support a request for six fire fighters per company. Be cautious when using this type of strategy; it can sometimes backfire.

If the consultant recommendations do not support the budget requests, then the chief probably has lost the battle. In most cases, the consultant's recommendations will be reported to the public by the press. Again, only use this strategy when you're absolutely sure that the request is unquestionably justified.

When managing a major capital improvement project such as a fire station, revenue generated by the sale of bonds will be used. The use of these funds has slightly different rules that the administration should be familiar with to keep out of trouble.

First, these funds can be used only for the capital improvement project and cannot be transferred to any other use. In some cases, dependent on local or state laws, funds may be able to be transferred from one project to another project as long as both are capital improvement projects. For example, at the end of a construction project for a new fire station, there may be some funds not needed from the bond sale. At this time, the department would like to start a rehabilitation project at another fire station. Although approval procedures vary, it should be possible to transfer these funds to this other capital improvement project. This can reduce or eliminate the need for a bond sale for the rehabilitation project.

One advantageous technique for major construction projects for new facilities is termed a *turnkey* project. Essentially, the project is funded to pay for the construction of a fire station, all its furnishings, and, in some cases, new fire apparatus for the station. However, fire fighters' salaries would not be able to be paid from the bond sales because these costs are part of the operating budget.

On the operating budget side, funding for additional fire-fighter positions to staff this new station must be budgeted separately. If this is a new volunteer station, the administration should start recruiting new members at the earliest opportunity. In many cases, for volunteer departments, it may be better to relocate a station because of the difficulty in recruiting new volunteers.

Fiduciary Audits

Periodically, the department will be audited for its use of public funds. Some items that the auditors will be looking for are misappropriation, embezzlement, accidental double payments, theft, and failure to follow procurement policies. The consequences of improprieties or negligence can include prosecution, loss of funding, loss of a trustworthy reputation, and additional regulations.

At some point in becoming familiar with a job that contains fiduciary responsibilities, the chief should find out about any random or scheduled audits. Check with the predecessor and other government financial officials and obtain a briefing on the last audit. This is one of those areas where it pays to be very detail oriented.

If there is an occasion when the purchasing policy was not followed and there was a good justification, make sure to thoroughly document the situation and all the details. For example, the department had an emergency situation that needed a large quantity of Class A foam for an unprecedented large wildfire. In most governments, there are procedures to follow for emergency purchases. However, in this case the chief bought from a vender not on the approved list and later the auditors asked for documentation of the unapproved purchases. Without documentation, would the chief be able to remember all the details and produce witnesses that can substantiate their account of the emergency situation?

Strategies to Counteract Adverse Economic Impact

On a continuous basis, analyze the local political climate and maintain close contact with elected officials and their administrative staffs. Also, seek out influential private citizens in the community and cultivate an ongoing friendly business relationship with as many of these important people as possible. In many cases, these private citizens have as much, if not greater, persuasive power over the decision-making process of local government than the elected officials.

Have facts and data ready. Statistics may be the best way to demonstrate the benefits of existing service levels and the impact of reductions. Identify and document all benefits to the community of current services.

If forced to cut the department's budget, have a plan that protects the highest priority items. It may be very tempting to choose items that are the least controversial to cut, but in the long run these types of cuts will have a distinctly adverse impact on service. For example, when faced with mandatory cuts in their budget, many chiefs of fire and emergency services organizations select training or prevention items for the cuts. This can be very short sighted. The long-term effect of cutting in these areas can be substantial. It may take a year or more, but cuts in training will impact the quality of services to the public.

Protecting the budget takes many years of preparation. The chief needs to build political alliances and sell the department's programs to key political officials and influential citizens. It may seem very unlikely that a financial crisis will affect the city or county, but it is always a possibility for which the chief cannot afford to be unprepared.

As any good salesperson knows, the department needs to be constantly selling its products and services. This includes educating the public. A cadre of speakers should be provided to accommodate any requests for presentations. It is also a good idea to offer presentations to the public, especially to public interest groups and organizations.

In good financial times, the budget will get only a cursory exam before an almost automatic approval. But when the budgets must be cut or the department would like a substantial increase, the administration will need all the support that can be assembled.

The chief should have an internal audit of the budget done to identify any items that could be deemed unnecessary or unjustified. Remember, in good financial times, these items will get very little scrutiny. When the elected officials and the budget officials go looking for places to cut the budget, if they find unjustified items, these items will be cut—and they will assume there are more unjustified expenditures. Therefore, if chiefs can eliminate these items ahead of any budget crisis, their reputation of being a professional public servant may help the department hold onto more of its budget during hard times.

In other words, make those justified changes in the organization and its budget before the crisis arrives. Because these are changes, remember to expect opposition. Everything in the budget has a supporter. The trust and confidence of the elected officials and the public are absolutely essential to winning the budget game. If nothing else, at least the department will be able to hold onto its fair share of the tax revenues and not see the budget raided to fund other public services such as police, public works, and schools.

Money is the lifeblood of a fire and emergency services department. If administrators can gain their fair share of the tax dollars, they have been successful at one of their most demanding and challenging duties. Nothing can be done without funding.

CASE STUDY: A Master at His Art (Page, 1998)

If you've got a copy of *America Burning*, take a look at pages iv and v. The eagle-eyed man pictured in the center died this February 28. Keith E. Klinger was one of the most effective leaders I have ever known.

After he retired from the fire department, Chief Klinger served as a member of the National Commission on Fire Prevention and Control. At the time, 12,000 American lives and more than $11 billion in resources were lost to fire each year. Twenty-five years ago this month, the commission delivered its report (*America Burning*) to President Nixon. For a variety of reasons, our national death toll and fire loss per year has been cut in half since then.

A few years before Chief Klinger retired, a fire started in Bel Air, California, when combustible vegetation ignited in the Santa Monica Mountains. Within 72 hours, nearly 600 homes had burned, and the mountains looked like a blackened moonscape.

On the first night of the fire, I was driving Assistant Chief Victor Petroff. We were on the move all night long, as the fire moved on a front several miles wide, eating its way through the brush, flaring up and dying down with each change in wind, slope, or fuel density. Several times during the night, we met on dusty fire roads with Chief Klinger and his driver.

When we weren't meeting with him, we could hear the boss on the radio, seeking updated information, arranging for meetings with his sector chiefs, inquiring about the welfare of personnel, and scolding food dispensers to "get over there and take care of those guys." He was 50 years old at the time, but he didn't slow down or sneak off for a nap all night long.

At about 7 a.m. the next day, Chief Klinger held a shift-change meeting with off-going and on-coming chief officers. He was wearing his dark blue work uniform and it was covered with soot and dust. He was energetic and forceful despite a night without sleep. As the meeting ended, Chief Klinger said to his driver, "Sully, take me downtown," and they drove off in his silver Olds sedan. The rest of the story was told to me by the late Kenneth Hahn, a member of the Los Angeles County Board of Supervisors (five of the most powerful elected officials in California). According to Hahn, the board was having its regular Tuesday morning meeting, and the news was all about the Bel Air Fire and homes lost.

"All of a sudden," Hahn said, "the wooden doors at the back of that big meeting room swung open, and through them marched Chief Klinger. He was covered with soot and dust, and I swear he must have had a fireman out in the lobby with a bellows full of smoke, puffing it through the doorway after the Chief. He marched down that aisle like he'd just bought the building," Supervisor Hahn continued. "He wasn't on our official agenda, but he walked right up to the podium and took over the meeting."

"What could we do?" Hahn asked rhetorically. "It seemed the whole county was on fire and the Fire Chief wanted to talk to us. Chief Klinger knew how and when to get attention."

"I noticed he had a folder in his hand," Kenny Hahn remembered. He then recalled how our Chief gave the board a blow-by-blow report on the battles that were underway in the mountains between Sepulveda Pass and Topanga Canyon. Then, Hahn recalled with a grin, Chief Klinger pulled from the folder a 10-year plan for improvement of fire protection. It called for spending millions of dollars. Again, Hahn asked, "What could we do but vote yes on it?"

The Board of Supervisors adopted Chief Klinger's 10-year plan and provided the money for it, including a fleet of specially designed brush-fire engines and several new fire stations in mountainous areas. Obviously, our chief had the plan developed long before the Bel Air fire and had been waiting for the best time to spring it on his elected bosses.

Kenny Hahn and the other supervisors knew they had been set up. But, judging by Hahn's jovial recollections of the event years later, they admired the performance of a fire chief who was as good a politician as any of them.

There were many other examples of Keith Klinger's power and effectiveness as a leader. When he retired, the supervisors praised him for the strength and independence he had brought to his position, but they replaced him with someone they'd be able to keep under their thumb.

There aren't many times in life when we get to see a first-class leader doing some of his best work. It was my good fortune to watch and learn from Keith Klinger up close and personal, and I'm one of many who will miss him.

Source: Reprinted by permission of JEMS and *Fire-Rescue* magazine

Discussion Questions

Read *A Master at His Art*, by Jim Page, and discuss the following questions:

1. Identify an existing need in a fire and emergency services department.
2. Quantify the resources, both staff and equipment, needed to implement the program.
3. Provide a list of all costs.
4. Provide an overview of the existing opposition to the program.
5. Describe a hypothetical or real situation that could support an emergency request to fund the program.

References

Bruce, Neil. *Public Finance and the American Economy*. Reading, MA: Addison-Wesley, 1998.

General Accounting Office (GAO). *Tax Gap: Many Actions Taken, but a Cohesive Compliance Strategy Needed*. GAO GGD-94-123. http://www.gao.gov/ (accessed August 6, 2005), 1994.

Institute for Policy Innovation (IPI). *Hidden Taxes: How Much Do You Really Pay?* by Bryan Riley, Eric V. Schlecht, and Dr. John Berthoud. http://www.ipi.org/ (accessed on August 6, 2005), 2001.

Internal Revenue Service. *Individual Income Tax Returns*. Washington, DC: Statistics of Income Division, Unpublished Statistics, 2002.

NFPA. *1710, Standard for the Organization and Deployment of Fire Suppression Operations, Emergency Medical Operations, and Special Operations to the Public by Career Fire Departments*. Quincy, MA: NFPA, 2001.

NFPA. *U.S. Fire Department Profile through 2003*. Quincy, MA: NFPA, 2004.

Page, James O. A Master at His Art. *Fire-Rescue Magazine*, May 1998.

Reuters News. *U.S. Business Economists Fear Stock Market 'Bubble.'* http://global.factiva.com.lp.hscl.ufl.edu/en/eSrch/ss_hl.asp (accessed August 6, 2005), August 18, 1999.

United States Fire Administration, FEMA. *Guide to Funding Alternatives for Fire & EMS Departments*. Emmitsburg, MD: FEMA, 1998.

United States Fire Administration, FEMA. *Recruitment and Retention in the Volunteer Fire Service*. Emmitsburg, MD: FEMA, 1998.

5

Human Resources Management

Knowledge Objectives

- Understand the importance of well-trained and well-equipped professional fire fighters.
- Understand the function and operation of human resources personnel.
- Recognize ways of providing diversity in the department.
- Examine the legal issues dealing with hiring fire fighters and recruiting volunteers.
- Comprehend and review the disciplinary process along with its legal problems.
- Examine the process of job analysis and validation.
- Discuss the influence and operation of public sector unions.

The Most Valuable Resource

The members of a fire and emergency services (FES) organization are always the most valuable resource in terms of costs and results (**FIGURE 5-1**). Without well-trained, well-equipped, and well-prepared members, the essential tasks at an emergency scene cannot be accomplished effectively, safely, or competently. The other issue critical to success on the emergency scene is equipment, consisting of stations, emergency apparatus, safety equipment, and specialized tools.

The cost of one engine company amortized for 1 year is:

- *Staff:* Five fire fighters (minimum crew of four plus one to cover leave impact) at $42,000 salary each plus fringe benefits (~40%) for three shifts (53 hours/week) = $882,000
- *Facilities:* One fire station ($3 million using a 30-year bond [6% interest] = $217,947 annually) and one engine ($300,000 using a

20-year bond [6% interest] = $26,115 annually) = $244,062 per year

The people costs for this engine company are more than three times as much as the capital items needed to respond to emergency incidents. And this does not include the *overhead* expenses of staff and administration personnel. Behind each piece of emergency equipment are people who need to be hired or recruited, trained, evaluated, and motivated to do the job well and professionally. Having individual members ready and able to do their job is what fire and emergency service personnel management is about.

Human Resources or Personnel Office

The human resources or personnel office is the agency within the local government that is responsible for recruiting, hiring, training, and paying employees. In theory, these agencies are support staff for the line agencies such as police, fire, and schools. Generally,

FIGURE 5-1 Alachua County (Florida) Fire Rescue Engine Company

merit rules and regulations along with union contracts are used to provide direction in carrying out the duties of the job. Merit rules are legally adopted by the jurisdiction and contain provisions for pay, benefits, discipline, and job descriptions for employees.

The employees of these agencies generally have specialized training and education; many have college degrees. However, their real understanding of fire and emergency services personnel issues will generally be limited. These employees can play a major role in acquiring additional funding for personnel and selecting the best people for various jobs.

"The success of a fire service personnel manager will often depend on a strong and viable relationship with the personnel office of the town, city, or county" (Edwards, 2000, p. 15). The first step is to familiarize human resources personnel with the nature of fire and emergency services operations, which are truly unique. The fire and emergency services organization needs new members who have the ability to learn complex skills and knowledge, have courage, are physically fit, do not mind getting extremely soiled and sweaty, can work in temperature extremes of both hot and cold, can tolerate working with or around critically injured or deceased victims, can operate in a family-like situation for up to 24 hours, and are team players.

If the human resources employees are not fully aware of the need for these unique abilities, they may unintentionally hire individuals who cannot perform the essential job functions when faced with other policy goals dictated by the city administration.

Educate these public officials by taking them to a fire and emergency services station or the training academy and have them experience the working environment (minus live fire) and job tasks. Do not just describe the tasks, but have them don the protective clothing and self-contained breathing apparatus (SCBA), and perform some job functions. Some of the same multimedia materials, lectures, and demonstrations that are used in public relations or risk reduction education programs can be very helpful.

The chief should attempt to have as much influence over the personnel management functions as possible. For smaller organizations, a fire and emergency services officer should be assigned part time to coordinate with the human resources office; a larger department should have permanent staff assigned full time. These officers should have some formal education in personnel management. In some departments a civilian employee with the appropriate guidance and knowledge from the chief can carry out this function.

Diversity

Complete diversity is about gender, race, age, education, country of origin, and many other items such as personal values. In the past, the typical fire and emergency services organization was made up of people with a lot of similar characteristics—those who were different were expected to change. In today's world, each member wants and deserves to be valued as a distinct individual.

Most fire and emergency services organizations have actively recruited minorities and women to compensate for memberships that had been predominately white and male. In some cases, these organizations were under court orders or consent decrees to provide affirmative action programs to increase the numbers of minorities and/or women.

There is some disagreement about the use of affirmative action, preferences, set-asides, and quotas. These techniques have been used to hire minorities and women in many fire and emergency services organizations. This is one area where the law has changed as a result of new legislation and contemporary court rulings. For example, discrimination cannot be proven solely based on percentages of under-represented groups in the fire and emergency services organization's workforce. It was once common practice for judges to decide discrimination cases based on a lack of a specific percentage of the under-represented group as compared to the general population served.

Starting in the 1970s, coalition groups began to support the rights of particular categories of people. These groups are very vocal in their identification of perceived or real discrimination and demands for corrections.

Many different types of people are members of the typical fire and emergency services organization. Just because people are different does not mean there is a right or wrong set of values, only that they are different.

The University of Michigan's Affirmative Action Case

A 2003 Supreme Court case upholding affirmative action policy at the University of Michigan's (UM) law school was a surprise to those who believe affirmative action should not be used routinely to provide diversity. What makes this issue even more confusing is that the Court also ruled that the undergraduate policy at UM was not legally correct and must be changed. "In that admissions process, which the university now has to revamp, the school literally gave extra points for applicants who were black, Hispanic or Native American" (Vlahos, 2003). As a matter of fact, the diversity points could outweigh points given for academic achievement.

Some are arguing that this may lead to widespread lying on applications to schools and jobs. For example:

> Affirmative action-related fraud does have precedent. In 1990, the Boston Fire Department found that six of its firefighters had falsely claimed minority status on their applications, including white twin brothers Phillip and Paul Malone.
>
> Initially, the brothers failed to qualify due to low test scores on their first test. On new applications, they claimed a great-grandmother was black. Though they earned the same test scores the second time around, they qualified for jobs under new minority outreach standards. After 10 years on the job, they were fired for fraud. (Vlahos, 2003)

So what does all this mean for your department? The following hiring scenarios seem to comply with this latest Supreme Court decision and others:

- A policy that rates each applicant using scores or ratings on written and other exams and then chooses the person with the highest score first, and then others strictly by their ranking on a list. The Supreme Court did not say that this traditional way of selecting new fire fighters cannot be used legally, but it must use job-related criteria and must not discriminate.
- If there has been a court order that the institution has discriminated and must provide an affirmative action program, which would result in a mandatory plan and quotas, the court order must be followed.
- There is a compromise policy called *banding* that takes into consideration both points of view. In a banding program, each applicant is put into a band and this list of applicants must be exhausted before going to the next band. One common cutoff for the top band is 85%. After the applicants take their written and any other exams, they are put on a traditional list and ranked from highest to lowest. Then new members are selected from the top band first.
- In actual practice, the department may, for example, offer jobs to women or minorities in the band and then offer jobs to white males in a ratio that achieves diversity. This helps prevent what some would characterize as *reverse discrimination*. Some affirmative action programs have this undesirable feature.
- Banding with a cutoff of 85%–100% uses a rationale that is identical to the typical grading schemes in school. For example, an "A" grade is actually a numerical grade of 90%–100%. This takes into account the fact that human beings are consistently inconsistent in test taking. A statistician can prove that people have a *variance* in their test taking performance, which means that on any particular day, the numerical grade received and the position on a list can change for each individual within a predicted range. For example, one day's performance may see a person at the number 1 spot, but another day they may be 13th.

Diversity Selection in Practice

To make banding work well for diversification, it is necessary to normally not offer jobs to everyone in the band because the mix in the band probably is not consistent with policy goals. Therefore, for example, if there are 100 people in the band, then selecting 20–30 will probably provide a diverse group. To assure this outcome, the department will have to recruit and encourage minorities and women to apply for the positions. This is normally a job handled by an affirmative action officer in the department.

However, if the department has vacant positions and the ideal diverse mix cannot be achieved, the positions will have to be filled. There are numerous and varied situations in which this may occur, such as with smaller departments that typically have fewer applicants. The administration must try its best to meet the goal, and if it cannot, which is not unusual, the positions must be filled with qualified applicants.

It is never a good idea to lower the job requirements to select applicants outside the band, or to select applicants who cannot meet minimum requirements such as physical fitness standards. There are three reasons to turn down applicants who do not meet minimum job requirements:

1. Firefighting is a very hazardous job; skills and physical strength can be a matter of life or death.
2. If fire fighters are not able to adequately perform essential job functions, the public will receive inadequate service.
3. It would be confusing to the fire fighters who met the requirements, and they may need to make up for the incompetence of others.

Diversity Sensitivity Training

One fire and emergency services department started an affirmative action program that recruited minorities to apply for the fire-fighter positions. The affirmative action officer decided to go to some local churches where the membership was mostly minorities. This officer was able to find many applicants who successfully passed the written and physical fitness examinations.

These new fire fighters were assigned to permanent fire stations after completing recruit training. In the typical fire station, the crew becomes almost a second family for the fire fighter. Although there are always differences in personal values, many share or tolerate differences in values of crewmembers. Because they were recruited in churches, these new fire fighters had some strong religious values and personal beliefs that in many cases were not shared by some older members of the crews. For most of the shift, there was a distinct lack of socializing.

Many agencies have started offering sensitivity training that educates employees to be aware that there are differences that have to be taken into consideration for people to get along with each other. One book used in training, *Men Are from Mars, Women Are from Venus* by John Gray (1992), describes differences in the way men and women feel about actions and communication.

Sensitivity training can go a long way in melding these diverse personalities to work together in harmony. For some minorities and women, this can make the difference between feeling welcome and feeling alienated. If an individual feels alienated, he or she could become less motivated on the job, and may show evidence of carelessness resulting in accidents. In addition, the individual may resign earlier than planned as a result of feeling out of place, which becomes a big loss to the organization that spent many hours training the employee.

Generally, diversity should be accepted and encouraged. Society shows that different cultures can peacefully coexist, and the workplace should attempt to reflect that as much as possible. The public can more easily relate to a workforce that is representative of their society. Furthermore, diversity will provide a broader perspective in organizational problem solving.

While attempting to achieve this diversity, the department should not lower standards of performance. Acceptance of the new diversity of employees must be supported by minimum standards for job competencies. For these people to become part of the family, they must be able to perform the minimum job requirements. In the emergency services profession, members must rely on each other to help get the job done in teams and, more importantly, if one member needs help, any other member must be able to come to their aid.

It is unethical to hire or accept members into fire and emergency services organizations who are not capable of performing critical job competencies. The public trusts that fire and emergency services personnel can perform emergency functions safely and efficiently. If fire fighters cannot perform to professional standards, they will fail the public and fellow members.

It should be noted that many fire and emergency services organizations, in their efforts to *do the right thing*, have hired or accepted members who cannot meet minimum job performance criteria. These situations will be very challenging to resolve, requiring an administrator with a good understanding of leading change (see Chapter 3).

Recruitment and Selection

It is a good idea to assure that there is commitment to provide diversity in the organization as new employees are hired or new volunteer members are accepted. It is rare that a formal quota system can be legally used to provide the right mix. These are allowed only when a court has ordered a remedy to a proven past discrimination practice.

Therefore, the department has to set up a recruitment and selection process that attracts a diverse mix of applicants. There are many techniques for accomplishing this, but only the ones that are ethical and most easily implemented will be discussed here. The chief should consult with the human resources office staff for more options.

Because minorities and women tend to be under-represented in most fire and emergency services organizations, assign a member the responsibility of recruiting minorities and women to apply for jobs or membership. The member selected should be committed to making change and providing diversity.

A professional presentation should be created or acquired that will help educate and attract applicants. Then identify locations or specific communication channels to reach the selected audience. For example, visits to women's athletic events at local universities can turn up some good prospects who are physically qualified to be members.

Fire and Emergency Services Workforce Issues

A common workforce issue concerns the understanding that the typical fire and emergency services organization is different from other occupations. Most of those differences are due to the nature of the fire and emergency services work that is extremely time-sensitive, risk-oriented, and team-based, with members in close contact with one another for long periods of time. The duration of the typical fire and emergency services shift and the closeness of the club-like surroundings of the career and volunteer fire and emergency services organization make it imperative to manage diversity effectively:

> Fire fighters work, eat, and even sleep together.... Even law enforcement does not generally entail such a close working relationship; in fact, many police officers spend most of their day on patrol by themselves. This close relationship, sometimes called the *fire service family*, has the potential to greatly magnify differences in people, including the differences often found between races, genders and cultures. (Edwards, 2000, p. 27)

Legal Issues

This section will contain some up-to-date information from recent Supreme Court and legislative actions (**FIGURE 5-2**). The law is constantly changing. Many of the recent Supreme Court cases have been decided by a slim majority (5–4), so replacements for retiring justices may swing the Court in a different direction.

FIGURE 5-2 The Supreme Court

There are two aspects of legal issues that affect fire and emergency services: the recruitment and retention of members and emergency operations. This chapter will discuss only personnel management issues. Also, keep in mind that in the past ". . . few people would even think of suing the fire department as the result of their actions [emergency operations or human resource issues]. This is not the case today, nor will it be in the future" (Edwards, 2000, p. 35).

Please check with your legal representative for any changes in local, state, or national laws or interpretations that may make the legal opinions in this section inaccurate.

Hiring Issues

One of the most important labor functions is the hiring of new personnel or the acceptance of a new volunteer member. A mistake in this process can leave the department with an undesirable member that must either be kept or be separated. This is never an easy or pleasant process.

In the screening process, be sure to screen for skills, knowledge, and physical abilities that will be needed to master essential job performance functions. For fire fighters, the necessary skills for most job functions are not mastered until recruit training is completed.

It is essential to test for the basic skills needed to master the training program for new fire and emergency services members. Tests that indicate a minimum ability in reading, writing, math, and cognitive skills should be used. Psychological testing is another area that deserves consideration in light of incidents of destructive behavior and arsons by fire and emergency services members. Some basic testing categories that should be used are as follows:

- Psychological tests
- Aptitude and achievement tests
- Personality tests
- Integrity/honesty tests
- Physical fitness tests

To reduce the likelihood of liability, fire and emergency services organizations should use a cutoff score on screening exams that indicates the minimum qualifications necessary for successful job performance. To compensate for normal variances in all testing results and to guarantee that the applicants selected can perform at minimum levels of competency, both physically and mentally, over a 20- to 30-year career, the minimum cutoff should be above the theoretical lowest scores. It is not a good idea to hire or accept marginally qualified applicants; it is not fair to the applicants, the fellow members, or the public they will serve.

Reference Checks

Many companies have adopted a strict name, rank, and serial number approach to requests for information about present and past employees. This neutral reference information has backfired in some cases, causing liability exposure for these organizations.

Many companies have been reluctant to give any negative information about employees' performance and disciplinary employment records. There is now increased risk of negligent hiring and retention of employees, especially in organizations with public trust.

For example, a new employee was hired as a paramedic after a complete background check. A past employer failed to provide information that the individual had stolen jewelry at the scene of a medical emergency incident. The previous employer could be held liable for withholding information that was relevant to the employee's behavior on the job.

Florida recently enacted a statute that provides legal protection to employers who furnish information about present or former employees. This provides immunity from civil liability when sharing information about employees. Check with the state attorney general for a similar statute in your state. The employer is covered unless it can be shown that the information was knowingly false or deliberately misleading. In those cases, it would be a violation of the employee's civil rights and appropriate legal action could be sought.

The following are some guidelines for providing reference information:

- Designate one individual in the personnel department as the only contact for requests.
- Require that a written request be made on company letterhead. Phone the requesting agency to confirm all requests.
- Disclose only documented job performance information, not subjective evaluations.
- Do not disclose any information regarding discrimination complaints or medical/disability information that may reflect on Equal Employment Opportunity (EEO) status or a protective category.

The Financial Impact of Lawsuits

A study released in 1999 by Jury Verdict Research points out the following growing risks employers are faced with:

- From 1994–1997, punitive damages were awarded in more employment disputes and employer negligence cases than in any other type of lawsuit.
- Median jury awards between 1996 and 1997 increased 286% from $64,750 to $250,000.
- Punitive damages were awarded, in addition to compensatory damages, in 34% of discrimination cases and 38% of sexual harassment cases.
- The average compensatory damages awarded to victims of discrimination or sexual harassment increased from $188,347 to $212,598.
- Plaintiffs are more likely to prevail in discrimination and harassment suits than are defendants. In 1996, 51% of plaintiffs were victorious; in 1997 plaintiffs prevailed in 58% of these types of cases, indicating an upward trend.
- Recoveries by former employees are also significant. Awards to plaintiffs in constructive discharges and wrongful terminations averaged $461,745.

Recent Supreme Court Cases

Sutton v. United Air Lines, U.S. No. 97-1943 (1999) This case concerned two pilots who were refused employment by United Airlines even though both met the FAA vision requirements for the position, but failed to meet United's more stringent requirements. The two pilots had uncorrected vision of 20/200 and had corrected vision identical to United's more stringent requirements. The Court held that they were not disabled under the Americans with Disabilities Act (ADA), did not find the pilots substantially limited in any major life activity, and were not regarded as disabled.

This has application for those departments using medical exams such as the National Fire Protection Association (NFPA) Standard 1582, *Standard on Comprehensive Occupational Medical Program for Fire Departments*, which contains several items that a potential member may argue allow them to have accommodation under ADA. Remember, as long as the department's medical and physical requirements are based on job performance standards and are essential to the job, they can be used even if they discriminate.

Also of major interest is the Court's support of United's more stringent requirements. Remember, the NFPA standards are minimum requirements.

Kimel et al. v. Florida Board of Regents, U.S. No. 98-791 (2000)
The Supreme Court found that the Age Discrimination in Employment Act (ADEA) is not enforceable in state and local governments. Therefore, unless there is a state law prohibiting age discrimination (and some have an exemption for police and fire), a fire and emergency services organization would be allowed to have a mandatory minimum and/or maximum hiring and retirement age. For example, the federal government requires its fire fighters to retire at 57 years old. Because of this requirement, and because their retirement plan requires 20 years of service, the maximum hiring age is 37. In another example, a metro department had a maximum age of 29 for hiring and a mandatory retirement age of 55 years old. These types of age requirements were viewed as discriminatory in the past, but are now allowed in state and local governments.

If the department is using a mandatory retirement age, the cutoff age should be justified using hard data and studies. For example, the 2004 *Fire Fighter Fatalities* report from the NFPA shows that "The rate for firefighters [deaths on the job] in their fifties is two thirds higher than the average and for firefighters age 60 and over, it is 3.5 times the average" (NFPA, 2005b). This would seem to support a mandatory retirement in the 55–60 age range.

In explaining its decision, the Court stated, "Old age does not define a discrete and insular minority because all persons, if they live out their normal life spans, will experience it" (USSC, 2000, IV C).

Also, a yearly extension based on a comprehensive medical exam and physical fitness test could be allowed. The NFPA statistics have not been normalized for items such as medical and physical fitness levels. Therefore, it is likely that if fire fighters in good medical and physical fitness were studied, they probably would have different death rates than the general firefighter population.

In fact, NFPA reports that "Of these 308 firefighters, 134 (43.5 percent) had prior known heart-related conditions, such as previous heart attacks or had undergone bypass surgery or angioplasty/stent placement. Another 97 of the victims (31.5 percent) had arteriosclerotic heart disease, defined as arterial occlusion of at least 50 percent. This is a detectable condition, but the victims may not have been tested, and may never have been aware of their condition" (NFPA, 2005a, p. 2). Only 25% had no previous known or unknown medical indication of any heart problems. It appears that most heart attack victims had medical problems that should have resulted in treatment or forced retirement before the death occurred.

Civil Rights

Several acts of Congress have been based on the 13th and 14th Amendments to the U.S. Constitution, which guarantee equal treatment or protection against discrimination based on race, religion, gender, or national origin; these acts are generally referred to as civil rights acts. In the fire and emergency services profession, violations of these federal laws have resulted in monetary awards and court-ordered remedies typically labeled *affirmative action plans*.

This area of law has been slowly evolving and is different today than it was 10 or 20 years ago. This is also a complex subject, and legal advice should be obtained when considering a change in policy. Briefly, the following are a layman's firehouse lawyer summaries of the present situation:

- A statistical imbalance between the percentage of a protected class in the workplace and in the general population is an indicator of a possible problem that must be looked at carefully to see if any discrimination has taken place. In the past, this type of imbalance was prima facie evidence that there had been discrimination. Now it is only a possible symptom, not conclusive evidence.
- As a policy decision, if a department enters into an affirmative action plan (in written or not) voluntarily without a court order, and that plan contains percentage goals to hire minorities and women, the department may be open to civil court action based on reverse discrimination. Discrimination against any group is a risky business.
- Affirmative action plans can be used if they are designed to encourage and attract underrepresented groups to apply for employment/membership or pursue promotional opportunities.

Family Leave

The Family Medical Leave Act (FMLA) of 1993 has had an impact on many fire and emergency services agencies. It is not uncommon to have to hire additional personnel or provide overtime for leave impact when fire and emergency services companies have minimum staffing levels. Studies have shown that up to 20% additional personnel must be hired to compensate for leave in traditional sick, annual, and injuries on the job (IOJ) leave provisions. Without overtime, to maintain four people on duty, the department needs to hire five fire fighters.

The FMLA mandates that employers with more than 50 employees must offer employees up to 12 weeks of unpaid leave per year for family responsibilities such as the birth of a child; care of a seriously ill child, spouse, or parent; or the employee's serious illness. The employer must give workers their previous position back when they return to work. Therefore, during the absence, the fire and emergency services agency may have to cover the vacant position with overtime. Unless the employee is on some type of paid leave, the cost will be the difference between the salary of the absent employee and the overtime costs, which will generally run about 50% more. Larger organizations may hire additional personnel to cover the vacancies.

The additional cost to cover the shift with overtime may actually be a lot less than the additional 50% needed for overtime. A fire fighter's normal pay includes a base salary and benefits. It is not unusual to have benefits run up to 40% or more of the salary. Therefore, it will cost the difference between regular pay and overtime, or about 10%–15% more for each shift. However, if the fire fighter is on paid sick or annual leave for family leave, then the costs are 150% more. Clearly this can be a substantial unexpected cost.

If an employee needs IOJ leave to recover from a worker's compensation injury or is granted sick or annual leave for family obligations, the employer should put the employee on notice that the leave of absence is counting against FMLA leave. If this is not done, then once the employee is found fit for duty from their worker's compensation injury or runs out of sick/annual leave, the employee may be able to apply for an additional 12 weeks off, unpaid.

In addition, the employer should have a written policy that combines FMLA leave with other paid leave when appropriate. For example, if an employee is off on sick leave for a serious medical condition, the FMLA leave can run concurrently. It is important to have this type of policy to reduce employees' time off potential.

Injuries on the Job

Policies should be created to assure documentation and verification of injuries that are reported as *on the job*. Show genuine concern for the employee. Visit the employee in the hospital or make personal visits to the home. This should be done by high-ranking fire and emergency services officials as well as the employee's immediate supervisors, especially in cases of serious IOJ events.

Employees do heal faster if they feel like they are valued and missed. Send a get-well card and make appropriate visits during recuperation. Help the employee with the paperwork necessary for the workers' compensation claims and the payment of medical bills.

Stay in close contact with the IOJ members until they are released to either *light duty* or *full duty*. Keep a constant dialogue with the employee to inform him or her of what's going on at work. If the department's official cannot get there in person all the time, send notes often. And finally, plan a small welcome back gathering with the employee's close peers. Something as small as a continental breakfast can be very effective.

Light Duty

Light-duty positions do not have to be offered to an employee who cannot perform the essential functions of their job. For example, suppose an employee is injured on the job and is covered by workers' compensation. After a recovery period, the medical evaluation is that this person will never be able to fully recover to perform the essential functions of a fire fighter. It is not uncommon for the employee to want to work in another position that does not require strenuous physical work, such as in the dispatch office. However, this would require the department to use one full-time fire-fighter position from the staffing of fire/EMS units.

Some employees have attempted to use ADA and its accommodation feature to claim that the employer has a responsibility to create a light-duty job in the fire and emergency services organization. This is not a requirement of ADA.

However, the employer may offer a job in another agency that the employee would be able to manage with their permanent limitations. As far as the original job is concerned, the same rules as a new hire apply. The employee must be able to perform the essential job functions.

To lower workers' compensation costs, fire and emergency services departments will often bring individuals back to a light-duty position if they are still recovering and are not ready for full duty. The temporary duty assignment should be a meaningful job and not work such as answering phones or filing. For those fire fighters who are exemplary members, temporary assignments can be provided in positions such as fire inspections, fire safety education, or 9-1-1 dispatching.

While on light duty, provide the member with written guidelines covering work hours, tardiness/absenteeism, rest periods, and personal appearance. Specify the procedure for requesting leave including IOJ leave for medical appointments and physical therapy.

Providing a light-duty job for exemplary employees is a good policy because the longer employees are off the less likely they will return to work. This is what some refer to as the *soap opera syndrome*. Once employees get used to a sedentary lifestyle, some lose the motivation to come back to work.

This technique seems especially effective for employees who are on shift work. If the administrator brings the injured worker back on light-duty day work, 9 to 5, there seems to be a strong correlation to the injured employee wanting to return to full duty faster than expected.

This policy is effective even when the department can bring the employee back to duty only for short periods of time at the beginning. For example, only mornings with the afternoon for physical therapy. The important point is to get them in uniform and back to work as soon as possible.

Drug and Alcohol Testing

The fire and emergency services organization should have a written policy prohibiting members from using, consuming, or being impaired by drugs (legal and illegal) or alcohol. When the policy is first implemented, all existing members and any new applicants should sign a consent form for testing that would be implemented in the following circumstances:
- Pre-employment or pre-membership
- Random
- Post accident or injury (ASAP)
- Suspicion based

If a member refuses a drug or alcohol test, produce a copy of the signed consent form and remind the member that dismissal will result if they reject the test. The alcohol test could use the pass/fail criteria allowed for a determination of impaired driving under the state's motor vehicle statutes. Refusal to submit to a drug/alcohol test or a positive result should be considered legal grounds for immediate dismissal.

Sexual Harassment

There are two types of sexual harassment that the administrator should be aware of: quid pro quo and hostile work environment. Both of these expose an employer to substantial liability that can result in large monetary awards. Therefore, it is important for an employer to set up policies, training, and procedures to prevent and eliminate this conduct in the rare instances when in occurs.

These cases can be very complex, and courts have found that a consensual sexual relationship is not necessarily a welcome one. One key to whether there is sexual harassment revolves around the issue of whether the sexual advances were welcomed or not. For example, at work a member becomes friendly with another employee, and eventually asks that person for a date. At that time, there is no knowledge of whether the advances are welcome. If turned down for a date and it is made clear that there is no romantic interest by the other person, then any future request for a date may constitute sexual harassment. This is especially true if the person proposing the date is the supervisor of the other employee.

Hostile environment refers to employees' conduct of a sexual nature that may be offensive or intimidating. The simplest example of a hostile environment is allowing male employees to post pictures of women either naked or in sexually implied poses. The telling of sexually explicit jokes is another good example.

The Equal Employment Opportunity Commission (EEOC) has outlined the following factors for determining if a hostile environment is present:
- Whether the conduct was verbal or physical or both
- How frequently it was repeated
- Whether the conduct was hostile or patently offensive
- Whether the alleged harasser was a co-worker or supervisor
- Whether others joined in perpetrating the harassment
- Whether the harassment was directed at more than one individual

The following are some guidelines for actions that will be helpful in preventing and defending the organization in sexual harassment situations:
- Implement a written policy of anti-harassment with examples of prohibited actions.
- Provide training for all members including the kind of conduct that is illegal or that is not acceptable.

- Have all members sign a receipt that they have taken the training and understand the policy.
- Clearly post the policy in work areas.
- Identify the persons or office to be contacted.
- Encourage members to report any potential situations that may lead to a violation.
- Vigorously investigate and discipline any violations.

Dating Policies

It can be very difficult to recognize whether actions such as sexual propositions, sex-based comments, or leering and staring are welcomed or not, so some organizations have created formal policies covering these potential relationships at work. An outright ban has been found to be neither possible nor preferable.

In a fire and emergency services organization where employees are assigned to a shift crew, it may be preferable to have a policy that requires the transfer of one of the partners in a consensual relationship. This also solves the problem of a possible quid pro quo situation where a supervisor is one of the partners. In the typical fire and emergency services station, employees spend long shifts together and sleep at the employer's facility. A different policy can be used for staff employees who work in an office.

Even in a volunteer department, dating can be disruptive to the relationships of the other members if there is any hint of favoritism. A strong chief and an oversight board can be very effective at preventing any adverse consequences.

The Americans with Disabilities Act (ADA)

This act provides civil rights protection to people with disabilities and guarantees equal opportunity in employment, public accommodations, transportation, government services, and telecommunications. Because of the intimidating nature of a federal regulation and the different analyses by many lawyers, some fire and emergency services organizations have hired or accepted new members who physically or medically are not capable of performing the job of fire fighter or emergency medical attendant. Be very careful when getting advice about the ADA. Research the many articles that have been written specifically about fire and emergency services workers.

Very simply, if standards for job performance are based on verifiable requirements, then they are acceptable and legally defensibly. Even though job standards may eliminate some people with disabilities, the standards are valid and meet the exceptions stated in the ADA.

This act is designed to protect individuals from discrimination based on a disability that substantially limits a major life activity. The EEOC enforces the ADA. The employer must make a *reasonable accommodation* for a mental or physical disability of the individual, unless it would pose an *undue hardship* on the employer's operation.

In most cases to date, reasonable accommodation for disabilities has been shown to create undue hardship if applied to public safety jobs. Therefore, it is very important to document job performance requirements (JPRs) along with the medical, physical fitness, and mental abilities needed to perform the JPRs.

Do not fall into the trap of lowering the minimum requirements for medical, physical fitness, or mental abilities to accommodate a perceived exposure to legal action based on the ADA, civil rights, or another federal law designed to protect against discrimination. The goal of all fire and emergency services organizations is to give the best service possible within financial and practical limits of human performance, both mental and physical. Remember in *Sutton v. United Air Lines*, U.S. No. 97-1943 (1999), the Supreme Court sided with United Airlines' more stringent vision requirements.

The one issue that has not been completely resolved is physical fitness. At the present time, there is no NFPA standard on physical fitness performance for fire fighters. This is a complicated subject. There are consulting firms who can help devise a physical fitness program for the department. The U.S. Fire Administration (USFA), International Association of Fire Chiefs (IAFC), and International Association of Fire Fighters (IAFF) also have some good resources. Use all for a thorough and impartial review of the issues.

The ADA and Hiring Practices

In the hiring process, applicants cannot be given a medical exam or asked any specific questions concerning any medical or physical disabilities. However, applicants can be screened using mental and physical ability tests. These must be validated as testing for essential job functions. In addition, drug and alcohol testing is allowed prior to a job offer.

Once a job offer has been accepted, the new employee can be required to take a medical exam that may be used to rescind the job offer. In addition, it is very important, especially in fire and emergency services organizations, to have an extended probationary period during which the employee can be separated without cause. It may take many months to evaluate a fire fighter in actual emergency situations that may occur infrequently, such as a large structural fire. Also, it is typical and preferable to have a 1-year apprentice phase during which the new member learns additional knowledge and skills.

Any standards for separation should be validated for essential job functions and each employee should be made very familiar with the rules. Although legally employees have very few rights when they are probationary, it is better to make sure everyone, especially the probationary members, knows what the rules are, how they will be evaluated, and the justification for them.

The probationary period should run at least 1 year after the successful graduation from a recruit fire-fighter school. This will give the department a whole year to evaluate the new employees and be able to separate them if they do not meet minimum expectations.

Pregnancy Issues

If a pregnant employee cannot perform the essential functions of her job, should the department provide an alternative light-duty job? In most cases the answer would be yes, this would be a good idea. However, there is no absolute requirement to provide light-duty assignment. If the general practice is to provide other employees (especially males) who have temporary disabilities with light-duty positions, then it would be a good idea to do the same for a pregnant female.

Generally, the department can order a pregnant employee to take a job performance test or be evaluated by the department's physician at the beginning of the third trimester and earlier if there are any medical complications or observations of substandard job performance. Adopt a written policy with the input from the physician, the union, and the department's legal advisor.

A pregnancy creates extra weight, changes in equilibrium, and loss of agility. Any of these situations can cause injuries and failure to be able to complete essential job functions. This is another good example of why mandatory physical fitness testing for candidates and incumbents is so important.

Previously, employers have prevented pregnant workers from doing their jobs simply because of a concern for the well-being of the fetus. There is some evidence of medical harm to a fetus from elevated temperatures and the high concentration of carbon monoxide typically found in firefighting. However, the Supreme Court has held that the well-being of the fetus is only the concern of the woman.

Fair Labor Standards Act (FLSA)

The core requirements of the FLSA are fairly simple. Once covered, when employees work more than a designated number of hours in a pay period, they must be paid overtime. However, public employees have some special rules. First, the employee can accept compensatory time at a rate of $1\frac{1}{2}$ times instead of overtime. Also, fire fighters' work period is defined as a 28-day cycle and the average hours per week before overtime is a maximum of 53 (212 in 28 days). This was done to accommodate the common 24-hour on and 48-hour off shift that many fire fighters work.

This act has caused some controversy in two areas. First, the definition of fire fighter contained the provision that a majority of on-duty time be spent in fire protection duties. Some fire departments that also provided EMS typically assigned fire fighters to staff these units and they worked over 40 hours per week. This is the cutoff for overtime for all other covered FLSA workers (other than police, who can work up to 171 hours in a 28-day cycle [U.S. Department of Labor].

Fire fighters assigned to EMS duties full time filed several lawsuits and eventually won their cases. These were very costly to these jurisdictions, which had to fund large back pay awards. With more and more fire departments offering EMS, this issue became a major problem.

In response to this conflict, Congress passed a new law that now defines a fire fighter as including EMS, fire prevention, and other duties where life, property, or the environment is at risk. This now allows fire and emergency services departments to use the 53-hour workweek as long as these workers are dual trained, fire and EMS.

The other issue surfaced when it was noted that overtime could be awarded if employees volunteer in the municipality where they work. For example, in a large county fire department, an employee may work at Station 1 and be a volunteer member of Station 15. However, because of the overtime interpretation, paid employees have been instructed not to volunteer at any fire and emergency services station in their jurisdiction. This has caused some hardships in combination fire and emergency services departments.

In a recent appeals court decision, *Benshoff v. City of Virginia Beach*, 180 F3d 752 (USCA, 1999), the court dealt with the issue of Virginia Beach career fire fighters volunteering at independent volunteer rescue squads in their jurisdiction. These volunteer stations are completely staffed by volunteers and govern themselves. The rescue squad volunteers also did no firefighting. This is clearly a special case, and the court decided that the FLSA did not grant overtime to career fire fighters volunteering at these independent volunteer rescue squads. However, this decision holds only in the U.S. Fourth Circuit Court of Appeals: Maryland, Virginia, West Virginia, North Carolina, and South Carolina.

Insubordination

In an effort to improve morale and workplace harmony, employers may attempt to silence employees who regularly spread unsubstantiated rumors, vociferously challenge management, and publicly attack an organization's practices and policies. In this endeavor, the administration should be careful not to violate a person's First Amendment rights (freedom of speech).

The government has a legitimate interest in promoting the trust of the public in the services it provides. Therefore, unsubstantiated public statements can be handled as insubordination. Also, any high-ranking officer can be ordered not to communicate with the public.

The preferable way to handle this is to assign a public information officer (PIO) who normally gives all official statements to the media. Then, by written order, direct all members to refer any inquiries to the PIO. This prevents confusion by the public when an official statement is released.

Silencing Complaining Employees without Violating the Law

First and most important, psychologists recommend that the best method of reducing worker complaints is to create a safe, respectful, and productive workplace. The administration should listen to and take action, when justified, regarding the concerns and views of members. A formal process for employees to submit, anonymously if preferred, comments and constructive criticism should be provided. In today's age, use both a post office box and an e-mail account to receive this feedback. In addition, some departments use a telephone number connected to a voice message recorder.

It is also important that these concerns be acknowledged and that attempts be made to correct these unsafe situations. Some type of feedback should be given to employees, such as an employee newsletter indicating the department's position and any attempts to solve the situation. As long as the members see attempts to solve the problem, they will support the department's efforts.

For those complaints that are completely unfounded, the chief may want to follow this five-step procedure:

1. Remain objective, even in the realization that some members are chronic complainers and there is always a group that is opposed to any changes.
2. Focus on conduct when the complaining is disruptive in nature.
3. Determine if the member needs or could use some professional mental health services for truly disruptive behavior, and then privately suggest to the member that the employee assistance program can be helpful for their dissatisfaction. This should not be put in writing because the chief is normally not a mental health professional.
4. Evaluate observations and supporting evidence to see if the behavior calls for disciplinary action.
5. Select an appropriate level of discipline.

Public Sector Discipline

There is the general perception that once a public sector employee is hired and finishes probation it is impossible to terminate the person. This is not true! However, public sector employees have *property interests* in their jobs that require the employer to use due process in the disciplinary procedures to separate an employee. This is a legal term that indicates that rights to a job are similar to and as strong as the rights you have when you own a piece of real estate.

What does due process mean? Before an adverse action can be applied, the administration should follow these general guidelines:

1. Provide a written notice of charges, preferably a three-step process starting with a verbal reprimand unless the charge is of great consequence to the organization.
2. Include a complete explanation of the evidence and reasoning upon which the charge is based.
3. Provide a meaningful opportunity for the employee to be heard by an impartial decision maker, preferably with a representative of the employee, such as a union shop steward, present.

In addition, consider using the following checklist to help guide the administration through the separation or serious disciplinary action:

- Is there sufficient basis for discipline?
- Is there appropriate documentation?
- Is the inappropriate behavior related to conduct, such as insubordination or tardiness, or performance, such as failure to perform a job requirement?

- Was the employee given all their rights during the investigation?
- What is the employee's previous disciplinary record?
- Was the violation serious and were there any mitigating circumstances?
- Have other employees been disciplined for the same actions; if so, what were their consequences?
- Were all the appropriate steps in the disciplinary process followed?
- Did all the notices clearly detail the observed actions and the violation of the law or policy?
- Did the employee have adequate time, representation, and opportunity to respond to the charges?
- Was the agency's attorney consulted before proceeding with the notice of disciplinary action?

Fair, Reasonable, and Evenly Enforced Discipline

The first step in ensuring that discipline is fair is to make sure most of the potential inappropriate behaviors and performance standards are listed in written policies and standard operating procedures (SOPs). In some cases, a general statement and explanation may be sufficient, such as in the case of insubordination. In others, as in the situation of illegal drug and alcohol use while on duty, be specific. Generally, it is a good idea for the administration to keep these policies and SOPs to as few as possible. Remember, in the emergency services arena, it is always better to keep rules and regulations simple.

An ongoing training program to inform all members of inappropriate behaviors and minimum performance standards is mandatory. Members should be given adequate notice and time for compliance or appeals of the adverse action will in many cases be won by the member, allowing another chance and additional time to comply.

The department's disciplinary actions must also show a consistent and nondiscriminatory record of enforcement. For example, a department may have an SOP on driving that requires drivers responding to emergency incidents to come to a complete stop at all stop signs and red traffic lights. During the 3 years after the driving rules SOP was issued there was no attempt at enforcement. Then a driver was charged with failure to follow the department's SOP (this is the first written enforcement record) and a penalty of 1 month's pay is proposed. Between the severity of the adverse action and the lack of any prior history of enforcement, there will be difficulty sustaining this ac-

tion if appealed to an arbitrator or the courts. It is very important to have an active enforcement policy.

Also, the department should have a policy of *due process* that outlines a disciplinary procedure that starts with a written warning and, with continued violations, will result in substantial consequences. Appropriate notices of hearings and the ability to acquire legal or union representation need to be fair and realistic.

Probationary Period

A newly hired fire and emergency services employee should be put through a vigorous probationary program. Many potential employees present extensive qualifications, education, and experience to gain a job offer. This is a chance to gauge the effectiveness of previous education and experience.

Have a structured evaluation in place that contains all of the essential job performance criteria and standards. The assessments should be standardized at prescribed time periods including a formal written evaluation completed by the supervisor. The probationary period should emphasize feedback to the employee that will aid the individual in becoming a better employee, but also be designed to weed out those who are not qualified. A formal testing procedure for the basic essential skills and knowledge that the employee should have mastered in recruit training should be used.

A probationary member need not be told the reason for termination; the administration can simply state that the new member has not satisfactorily completed the probationary period. In most cases, it is better not to tell the probationary member the reason for termination, especially in writing. Although the employee has no winnable legal appeal, this may unintentionally encourage a lawsuit. Anyone can instigate a lawsuit, even if they have no chance of winning. Probationary fire fighters should be aware that they have not satisfactorily completed the basic fundamental skills or behavior expectations.

Develop a written performance evaluation that the supervisor can use and specify time frames for the completion. Train all supervisors in the use of the evaluation systems and encourage specific comments and observations in the report. Remind supervisors of the grave consequences of providing charitable probation evaluations.

Retaining a fire and emergency services member who is not qualified has double-edged consequences. The public will not be getting the best service and the member will not be able to operate safely, creating a hazard to themselves and the team.

Terminations

This section discusses methods for conducting terminations that lower exposure to legal challenges. There may be adequate evidence and support for a performance or disciplinary termination, but if handled inappropriately, it can lead to legal disputes that can be costly and disruptive to the fire and emergency services agency.

Members may be motivated to sue simply because they feel they were not treated fairly or with respect. The administration should approach the separation as a situation where there was not a good fit between the member and the job, rather than labeling the member as incompetent, lazy, or another derogatory characterization. This will help the member feel better about themselves. Not everyone has the aptitude or mental and physical fitness to perform the demanding fire and emergency services duties. They are not bad people; they are just in the wrong occupation.

On the fairness issue, everyone has a right to respond to their accusers and have competent representation in the administrative process leading to termination (except the probationary member). Even in cases where it may seem like a clear case, follow all the policies that provide due process for the member.

Reductions in force (RIF) are another situation that calls for terminations. It is absolutely imperative to have an established RIF policy and make all members aware of the selections process. What should be avoided are RIF decisions that may be discriminatory, such as selecting all members over 50 years old. From an accountant's perceptive, terminating one chief officer may be equivalent in salary and benefits to terminating two or three fire fighters. Form a task force of members to provide recommendations and adopt a written policy.

In the situation of a voluntary termination (the employee quits), gather as much information about the motivation of the member as possible. The objective is to uncover any issues that could later support a legal claim of discrimination by the member. Attempt to find out if there were any underlying reasons for the resignation that have not been previously stated by the member. Should these complaints have been addressed by the agency, such as would be appropriate for some type of discrimination? Ask the following questions:

- Why are they leaving?
- Would they consider returning to the organization?
- What plans do they have in the near future?
- Do they believe that the fire and emergency services should consider changes? If so, what changes and why?

Constructive Discharge

This is a legal term indicating that the member was somehow convinced to resign because of intolerable working conditions. Some common example scenarios are:

- The work environment was hostile (sexual harassment, etc.)
- Unsafe working conditions (check Occupational Safety and Health Administration [OSHA] and NFPA safety standards for compliance)
- Insufficient information about alternatives in disciplinary or performance actions
- Not given the option of continuing employment in lieu of an early retirement (sometimes used with RIFs to reduce the payroll)

State and Local Laws

Many states have statutes that reference fire and emergency services hiring and, in some cases, operations. For example, in Florida an applicant must be a state-certified fire fighter to become an employee of a municipality. The state specifies the curriculum using NFPA Fire Fighter I and II and a minimum number of hours (360) for the training programs. These programs are delivered throughout the state by community colleges, vocational tech centers, and a few larger fire and emergency services organizations. These training programs as well as their facilities are inspected and certified by the state. In addition, each graduate of these fire-fighter training programs must take a state-administered skills test and written exam for final certification. This provides a pool of certified fire fighters who become applicants for fire fighter positions.

At the local level there are many merit or personnel laws that regulate employment. In addition, union contracts are enforceable and in most cases have stipulations affecting salary and benefits. These are too numerous to discuss here, but are all legally enforceable.

Some states are known as *right to work* states because they do not provide for the organization of public unions. In these states, it is common for there not to be a union contract or formal discussions with the unions or other employee organizations.

Although this is not a chapter on administration, it is a good idea to meet with representatives of the employees on a regular basis and discuss topics such as salary, benefits, working conditions, and safety issues. This would be very good for the morale of the members.

Common Law

In the United States, many interpretations of existing laws and regulations become enforceable through the court system as a result of a previous court decision. Judges do not make laws, but they do interpret and apply them. These decisions become precedents that may be used by other judges in similar cases. This is called common law, and adds to the understanding and application of many laws and regulations. That is why it is important to keep up to date on every new precedent-setting court ruling. Remember that the lower appellate court decisions are only precedent setting in the federal district or, for state judges, the state in which they are issued. The exception to this is U.S. Supreme Court decisions, which are binding nationwide.

If the Supreme Court makes a decision based on the Constitution, this decision is permanent and can be changed only by changing the Constitution, which is difficult if not almost impossible. Although there is a process for adding to and changing the Constitution, it is used only in rare cases because it requires a three-fourths, super-majority vote by Congress and U.S. voters.

There is one big exception to this, however. Many of the Supreme Court's recent decisions have been made with a vote of 5–4. At some point in the future, as justices retire and are replaced, the Court may have a different opinion on some issues.

A recent example of this common law process was previously discussed in the section about the FLSA application to fire fighters performing EMS duties and their work hours. This case actually went to the Supreme Court and was decided in favor of the employees, granting them time and a half pay for all hours worked in excess of 40 hours. As a result of that decision, several national groups lobbied Congress to pass legislation to change the ruling by the Supreme Court. In record time, Congress did pass legislation that expanded the definition of fire fighter to include employees who function most of the time in EMS roles, but are cross-trained as fire fighters and work for a fire organization. This has the effect of these fire fighter/EMS employees having a 53-hour week as opposed to the 40-hour week for other employees providing EMS.

Finally, use caution in using reports of legal actions. It is not unusual for the administrator or someone else in the organization to read a relevant story in the newspaper or a magazine reporting a decision of a court or an out-of-court settlement. First, consult with an attorney and gain all the facts of the case. In addition, if the case was decided anywhere below the U.S. Supreme Court, it opens the door for appeals that may change the outcome.

Job Analysis

A job analysis is a common method of defining the requirements of a job. In fire and EMS, very detailed training and testing criteria are readily available. There are three distinct parts to a job analysis process: hiring criteria, recruit training and testing, and incumbent evaluation. All three should be consistent with the job analysis.

A good example of the outcome of a job analysis can be found on the following page.

At the present time, there have been very few legal challenges to many of the new employee requirements that most fire and emergency services organizations typically have in place. For example, a very detailed and comprehensive basic skills test is generally required of a new recruit.

The laws of this country require that all people be treated equally. However, incumbents are not tested periodically to assure that they can still perform the basic skills that a recruit must complete or be terminated. As explained earlier, a recruit on probation does not have the same rights of due process that an incumbent will possess. There may come a time when a future court decision requires equal treatment. Although this seems like it would be a fair and ethical policy to treat all fire fighters equally, it is common to have different requirements for incumbents.

Practically speaking, it would be a good idea to upgrade new employee and recruit requirements before upgrading incumbents, and then phase in requirements for incumbents. They will need time to relearn some skills.

Validation

Many human resources professionals recommend using incumbents, supervisors, and personnel analysts as sources to develop a job analysis. A good strategy includes some of each of these sources in the process.

This effort should start with a standard such as NFPA 1001, *Standard for Fire Fighter Professional Qualifications*. The administration should do some research and contact several fire and emergency services agencies to get copies of what is being used by peers. Ask if they have had any problems with their job descriptions. Are they happy with the abilities of their new recruits?

Once this basic research is finished, consolidate the information into one comprehensive job description. This draft should be circulated to the incumbents and supervisors in the department. In addition, actively solicit input from any labor organization and the personnel office.

SANDY HILL CITY

I. **Position Title:** Volunteer Fire Fighter
Revision Date: 10/2002
EEO Code: Protective Service
Status: Non-Exempt

II. **Summary Statement of Overall Purpose/Goal of Position:**
Under the direction of the Fire Chief and general supervision of the station officer; respond to emergencies involving fire, medical and environmental concerns. Responsible for the care, operation and condition of fire and rescue apparatus. Provides the City with prevention and mitigation of emergencies and disasters, through proper planning, public education and code enforcement.

III. **Essential Duties:**
- Respond promptly and efficiently to fire, rescue, hazardous materials and medical alarms.
- Operate at emergency fire and rescue incidents efficiently and safely.
- Maintain proficiency in fire and emergency medical knowledge and skills.
- Conduct tours, lectures and video presentations. Display fire apparatus. Participate in public demonstrations at local school programs and various civic and City functions.
- Inspect business, public and private properties for hazards and code violations.
- Drive and operate department apparatus in emergency and non-emergency situations.
- Maintain equipment on apparatus to include daily, monthly and annual testing. Test and rotate hose.
- Clean, wash, wax and repair apparatus.
- Complete daily, monthly and annual reports on the testing of fire and medical equipment.
- Mow, cut and trim station lawns. Vacuum and clean building. Remove snow. Ensure the building is ready for public inspection at all times.
- Maintain daily maintenance records and complete maintenance and equipment reports.
- Complete eight hours of department training each month.
- Must provide a minimum of three 12-hour shifts of service each month. These hours may be worked at any time with approval of the division chief or company captain.

IV. **Marginal Duties:**
- Maintain daily maintenance records and complete maintenance and equipment reports.

V. **Qualifications:**
Volunteer personnel must meet the following eligibility requirements:

At time of application screening:
- Must be at least 18 years old
- Must live within twenty (20) minutes of Sandy Hill
- Must not have been convicted of a crime for which the applicant could have been punished by imprisonment in the penitentiary of this or another state
- Must not have been convicted of an offense involving dishonesty, unlawful sexual conduct, physical violence, or the unlawful use, sale, or possession of a controlled substance
- The department attempts to schedule 4 persons on duty at the station at all times. Preference may be given to those applicants who are available during times with low availability of volunteers.

Before being hired:
- Complete entry-level written test
- Pass physical agility test

- Complete and pass drug test
- Successful driver's license check
- Check for convictions and outstanding warrants
- Volunteer agreement signed
- Establish HIV and Hepatitis B baselines
- Enter initial training program

Education: Valid State driver's license; State Fire Fighter I and First Responder certification required within one year of appointment. Must complete eight hours of department training each month.

Medical and Physical Fitness: Must meet department's medical and physical agility requirements.

Experience: Entry-level position.

Knowledge of and Use of: NFPA Fire Codes; fire department policies and procedures; fire ground procedures including fire streams, forcible entry, water supplies, salvage and overhaul; command procedures; inspection procedures; sprinkler systems; alarm and communication procedures.

Responsibility for: The use of discretion and judgment in emergency situations; the care, condition, and use of department apparatus, equipment and pumps. Respond to alarms when notified and available.

Communication Skills: Ability to communicate verbally and in writing; ability to write reports.

Tool, Machine, Equipment Operation: Frequent use of fire and medical equipment including hose lines, nozzles, pumps, hydrants, extinguishers, ladders, hand tools, extrication tools, air masks, etc.

VI. **Working Conditions:**

This job entails regular exposure to dangerous situations under disagreeable conditions, including smoke, heights, fire, fumes, heat, cold, emergency driving, etc.; frequent exposure to dangerous situations with medical emergencies; must meet Sandy Hill City Fire Department physical agility standards; must be able to wear and work in fire department breathing apparatus.

The above statements are intended to describe the general nature and level of work being performed by the person(s) assigned to this job. They are not intended to be an exhaustive list of all duties, responsibilities, and skills required of personnel so classified. The approved class specifications are not intended to and do not infer or create any employment, compensation, or contract rights to any person or persons. This updated job description supersedes prior descriptions for the same position. Management reserves the right to add or change duties at any time.

Because of the possibility that incumbents and their supervisors may not encounter critical essential job functions on a regular basis, they may not identify some of the job functions as being essential. Advise those questioned before doing their review that a critical job function may only have to be done once in a career. It becomes essential not because of daily use or need, but because it is how the public and the national standards expect a fire fighter to be able to perform.

To use an example from the ADA regulation, the expectation that a lifeguard at a pool must be able to bring a heavy person back to safety from deep-water depths is described in detail. Although this job function may happen only once in a career, it is one of the most essential because of the life-saving potential, and life saving is the ultimate purpose for a lifeguard.

The testing of incumbents can come in handy when planning an implementation strategy. This will provide a good idea of how many of the incumbents may have to improve to be able to successfully complete the tests. An implementation plan can be proposed that allows a phased-in compliance for existing members.

Job Classification

The greatest majority of members in any fire and emergency services organization are at the fire fighter level. The normal perception is that the officers and chiefs are more important than fire fighters, but this is not correct. The fire fighters are the people who get the job done at the emergency scene, and the organization should show its respect and acknowledgement of these special qualities, especially as each member gains additional training and experience. The department should continually highlight that the fire and emergency services profession is a noble pursuit that takes high levels of skill, courage, and physical ability.

Some fire and emergency services organizations have an annual awards ceremony that singles out individuals for heroic achievements. Most members are capable of heroic acts, but only a few fire fighters end up at the right place at the right time. Therefore, rewarding all members for their potential to perform at these heroic levels of service may be warranted.

To reward competent members, the fire and emergency services organization could use a three-tier promotion system for fire fighters, as follows:

1. Recruit Fire Fighter
2. Fire Fighter
3. Lead Fire Fighter

Each new level would have its own set of training requirements, testing, and experience on the job. For example, to gain the Fire Fighter level, the member would have to successfully complete recruit training (NFPA Fire Fighter I and II) and a 1-year probation program. Lead Fire Fighter would require an additional 3 years of experience, retaking of all of the basic skills and knowledge tests, and completion of an apprentice program. Each new level would be a promotion with an appropriate increase in salary.

Recruitment

In most cases, when a jurisdiction advertises a vacancy in the fire and emergency services organization, there are many applicants. "The challenge for most fire departments is to have a diverse pool of applicants that includes women and minorities who may be underrepresented in the current workforce of the fire department" (Edwards, 2000, p. 75). Edwards goes on to say, "Both equal employment opportunity and affirmative action requirements have dramatically affected the recruitment process."

In the past, it was simply a matter of choosing the *best of the best*. Now the question is, can highly qualified applicants be chosen that reflect the proper diversity mix? In any case, this should be one of the objectives of the department's recruitment efforts.

Always keep in mind that selection of new members is one of the most important decisions for the quality and effectiveness of the agency. Hiring decisions can and often do have long-range impacts regarding the competency and effectiveness of the department. People who do not meet the minimum job requirements should not be hired or accepted as members.

The Selection Process

Some of the legal ramifications of selecting new members were discussed previously in this chapter. Consideration should be given to cutoff scores and pass-fail requirements for cognitive abilities (speaking, reading, math, listening, and writing skills), medical tests, drug and alcohol tests, criminal background checks, physical fitness and strength, and cigarette smoking. For example, in some cases, fire-fighter job applicants must sign an affidavit that they have not used tobacco products for 1 year prior to being hired. Many departments follow this up with requiring fire fighters to sign a pre-employment agreement to not use tobacco at any time during their employment.

It is a good idea to set acceptable minimum scores high enough so that the department is selecting those who will be competent members in their ability to learn and perform the job. Selecting marginally functioning members will cause unacceptable performance in the future. Remember, these are selections for new fire and emergency services members for a minimum 20-year career.

Another issue is the option of hiring either previously trained and certified fire fighters or those not trained. Each option has it merits. The advantage of selecting previously trained fire fighters is that it reduces training costs to only a brief orientation. This is a substantial cost savings because a new fire-fighter recruit may be in training for 12–16 weeks. Also, many smaller departments cannot hire enough new fire fighters at one time to justify a rookie school. However, Edwards (2000, p. 79) points out "One of the major disadvantages of recruiting only trained and certified applicants is that the selection pool is much smaller. In many cases this pool consists of fire fighters switching departments for higher pay or benefits, or of volunteer fire fighters attempting to get paid positions. Because the demographics of the fire service in the United States are predominantly white male, this pool will consist of mostly white males."

Some prospective fire fighters may apply to a certified fire-fighter training center to become a student. The student pays his or her own tuition and other costs for the typically 12–16 week training programs. With successful completion of the Fire Fighter I/II course, applicants are eligible to take the state fire-fighter exam. Once they successfully pass the practical and written exams, they are state certified and can apply for a paid position in the state.

Some larger jurisdictions hire personnel and then pay them to attend the training programs. These departments prefer to train their own recruits so they can impart their organizational culture and the specifics of their department to the recruit fire fighters.

In addition, some departments also prefer paramedic certification. Many potential career fire fighters/medics must have both certifications to be selected for an opening. This trend is becoming more common throughout the nation.

Public Sector Unions

Unions are groups of workers that have formed an association to discuss issues with management. Many volunteer departments act like unions in that one of their priorities is to take care of the volunteer members. Unions protect jobs and attempt to improve the salary and benefits received from local governments. They also promote companionship and respect for fellow members. This creates a strong loyalty to the union or volunteer organization that sometimes can be in conflict with the main mission or goal of the fire and emergency services organization—namely, quality service to the public.

The relationship between unions (and volunteer organizations) and the representatives of the public (e.g., fire chiefs, mayors, county executive) is a checks and balances system—the administration is watching out for the best service to the public at a reasonable cost, and the unions are looking out for the rights of their members. This is an oversimplification, however; there are many situations in which unions and volunteer groups have supported changes that were primarily for the benefit of the public.

Unions come in all sizes and types. Some are extremely strong and influential; others are mere social clubs. In right to work states, unions tend to be weak and more advisory in their actions.

As an administrator, never underestimate the power of the local union or volunteer organization. You do not have to give in to all of their demands simple because they have influence; however, the chief must respect their opinions and give them a chance to voice their concerns and beliefs.

Strikes and Job Actions

"In general, public employees have no rights to withhold services from their fellow citizens, which forms the basis for the 'no strike' clause in federal and most state collective bargaining laws, especially with public agencies such as fire and police departments" (Edwards, 2000, p. 216). In most labor unions, the strike is the ultimate tool that can be used to gain agreement for an improved salary or benefit.

Without this negotiating tool, public safety unions are at a great disadvantage in discussions with the dominant governments. Some states allow binding arbitration for public safety employees. This permits a fair and equal status for labor and management. However, in right to work states, it is rare for there to be binding arbitration. In these situations, it is not uncommon for there not to be a written contract or for negotiations to continue endlessly. This is very bad for the morale of members, which may have a negative impact on service to the public.

Proposed U.S. legislation would allow collective bargaining for public safety officers, including fire fighters. The IAFF states, "A national collective bargaining law for fire fighters is the IAFF's number one priority. Currently, there are at least 18 states that prohibit fire fighters and other public safety officers from collectively bargaining over workplace issues like hours, wages, and working conditions" (IAFF, 2005).

Because this bill has just started its journey by being referred to committees in the U.S. House and Senate, it is unknown at this time (2005) if this bill will be enacted.

Bargaining Units

It is very common for there to be two bargaining units—one for fire fighters and one for officers. In many cases both units are represented by the same union president or are in some way combined into the same organization. This can cause conflicts in loyalties and discipline.

For example, if both the worker and supervisor are members of the same labor organization, they may become friends in the social atmosphere that surrounds these organizations. In some departments, only the fire chief is not a member of the union. When differences surface, it can become a one-person administration versus everyone else. This can isolate administration and make it very difficult to negotiate solutions to problems.

This policy has made it very difficult for supervisory or management personnel to separate their loyalty to the fire fighters and the duties they must perform as a supervisor. Both in the day-to-day running of the department and the negotiation process, supervisory personnel seem unclear on where their loyalty should be. This situation is exasperated by the nature of fire and emergency services work. Personnel perform their duties in small teams that become very close both professionally and personally. It is not uncommon for fire fighters to become very close friends.

Often, one of the friends is promoted and may become the others' supervisor. This may not be a problem of being part of the same bargaining unit, but it is one of being part of the same social organization (for both paid fire fighters and volunteers). Both labor and supervisors have several needs that are the same, including pay and working conditions. It could be argued by a union that this is clearly a case of equal goals and, therefore, supervisors should be part of the union.

One way to answer this predicament is to set up a policy that awards the same working conditions to both groups but insists on separate bargaining units. For example, when bargaining for pay raises, work hours, and other monetary benefits, the city could bargain with both groups with the understanding that the final outcome would be equal for both groups. For non-monetary or management prerogative items, separate discussions with each union president would suffice.

Finally, personnel who have middle management positions (above the company officer) should not be in any bargaining unit. These individuals should be a part of administration and be taken care of accordingly.

Local Government Representatives

When formal negotiations start, the management will be represented by several people from the government along with high-ranking officials from the fire and emergency services agency. These government officials may have a narrow point of view or goal, such as a target budget figure. They will not have the knowledge or experience to foresee any possible adverse effects on the emergency services that could result from a new union contract that saves the city money. This same lack of knowledge will be evident if using arbitration or fact-finding services.

An indoctrination program should be provided to inform these officials about the actual functioning aspects of fire and emergency services. For example, they may not understand why there is a need for four fire fighters per company when 61% of the department's calls are EMS that only require two people (a cardiac incident may require five or six). Emergency services are very technical and require a high level of skills and knowledge. Only when members have been providing emergency services for many years do they get an intuitive feel for what constitutes quality service, including the safety aspects of the profession.

Grievances

This is an important aspect of any labor agreement. This process allows for the fair resolution of misunderstandings of the contract or an appeal process for disciplinary actions. The union is obligated to represent the member to guarantee that fairness was used in the determination of any adverse action proposed.

This process also keeps many issues from being litigated in the courts. Most courts will not overturn a grievance procedure, especially if the process ends at binding arbitration. Therefore, when complying with the provisions in the labor contract for settling grievances, the decision by an arbitrator is normally final.

When processing a grievance, make sure the administration has all the facts and is well prepared for the hearing. If these facts indicate that the administration made an oversight or error, this should be corrected immediately. If not, include a full review and representation by the city/county attorney. The results of an arbitrated grievance can set a precedent that will be very difficult to change in the future. Do not take these cases lightly. They can have long-term effects on the department's operations and administration.

Progressive Labor Relations

It is beginning to be more common to see cordial relationships between labor and management. In the past, each looked at the other as the enemy and would do anything to win in negotiations and grievance proceedings. Nowadays, although each has to represent its own constituents, where there is a mutual interest, each is able to help the other out.

A lot of this mistrust was the result of confrontational behavior that was so common in the private sector during the Industrial Revolution. Even violence by management and labor was not uncommon. As a result of federal laws and regulations, and the revelation that to compete in the new global economy requires well-motivated workers, labor and management are finding ways to reasonably settle their differences without hurting the competitiveness of the industry or business.

This type of new cooperation is the result of improved communications. The chief should schedule meetings with union officials on a regular basis. In addition, visits to work locations to talk and listen to all members are very important.

There have been cases where the union or volunteer leaders have lost touch with their own people.

Their internal communications can break down. In these cases, labor and volunteer leaders may be using old paradigms or their own agendas as a basis for representing the membership. This is not a good situation. Unfortunately, it is relatively common.

Getting to Yes

This is the title of a very good book outlining a negotiating process that attempts to separate facts and issues from people and their emotions. "Principled negotiation methods focus on basic interests, mutually satisfying options, and fair standards that typically result in a wise and acceptable agreement" (Edwards, 2000, p. 237).

One great method to use in negotiations is to brainstorm for possible solutions. The ground rules should allow any suggestion to be recorded without making any judgments. Then the group needs to go back to each idea and analyze its cost, complexity, and reasonableness. Once a set of realistic possible solutions has been agreed upon, the separate parties can then concentrate on them.

The techniques described in *Getting to Yes* can also be used to problem solve differences between spouses, parents and their children, and many other interpersonal situations that find people not agreeing with each other.

Motivation

"Motivators affect the individual's sense of intrinsic satisfaction and provide a positive feeling toward work" (Edwards, 2000, p. 252). This is a very important concept for creating a satisfied member who will be a top performer for the organization. Some members can be troublemakers and can pull your entire organizations towards mediocrity if not dealt with by the administration.

To simplify this concept, the following two areas, when accomplished, seem to produce a motivated member: benefits (hygiene factors) and pride (motivators). The following are examples of benefits: fair salary and work hours, health insurance, workers' compensation, and length of service awards (volunteers) or pensions. The following are examples of pride factors: modern and well-maintained fire and rescue equipment, professional qualifications for membership, a rigorous training program, and, in general, anything that would lead a member to feel good about their contribution to the fire and emergency services organization and the organization's ability to provide a quality emergency service to the public. These factors can be reinforced by recognition from the organization such as positive reinforcement for doing a good job.

Pride factors seem to be more important to the individual. According to Frederick Herzberg, ". . . Motivators [pride factors] affect the individual's sense of intrinsic satisfaction and provide a positive feeling toward work" (Edwards, 2000, p. 252). But this is relative. The department may be able to have a well-motivated employee who is being paid less than the average for the immediate area, but not substantially less. Also, volunteers will serve without any pay if properly motivated.

Many of the pride motivators for volunteers can go counter to some contemporary beliefs. For example, some would argue that the additional time required for a volunteer to complete a fire-fighter comprehensive training program would be a disincentive to retaining members. However, Director Neil Good, in an article titled "Why Do Volunteers Volunteer?" in *IAFC On Scene,* had this to say: "Volunteers are attracted to the theoretical and science of fire fighting and medical. If the department has an active and rigorous training program, they will probably have a bunch of motivated volunteers. If retention is a problem in the department, first look at training program and secondly, look at the opportunity volunteers have to exercise their social values" (IAFC, 1995).

With a rigorous training program, loyal and motivated members, both volunteer and career, will be the norm. There also will be less disciplinary problems with these members who have a strong allegiance to the volunteer organization because they feel good about their ability to provide professional emergency service. Never underestimate a person's pride—it is a very strong human emotion.

CASE STUDY #1: Choose Your Own Case Study

In this exercise, students are asked to develop their own case study. First, use an Internet search to locate a recent appeals court or Supreme Court decision involving a fire, EMS, or emergency services issue. Then find as many articles as you can that discuss the decision in fire and emergency services periodicals. Summarize the articles by noting the specific details and points of law. Compare and contrast the articles. Once you think you have a good understanding of the issues and the court decision, find an attorney who will speak to you. Explain that it is a school project. Many attorneys will give you a limited amount of time without professional charges. A city, county, or state attorney would be another good source for expert advice. If possible, speak to more than one attorney because they sometimes have different views on a specific legal issue.

Finally, after writing a synopsis of the above, write a paragraph explaining your personal thoughts about the case. If you disagree with the decision, please explain why.

Limit your answer to no more than two pages.

CASE STUDY #2: New Officer

A newly promoted officer is assigned to a very busy engine company and on the first day is greeted with an energetic crew of three fire fighters who seem pleased with their new officer. In reviewing personnel files and contacting the previous officer, the new officer has acquired some background information on the crew and is assured that they are aggressive fire fighters who know how to put a fire out.

Before the morning coffee is finished, the company is dispatched to a reported fire in a single-family dwelling. The officer did not get a chance to explain any expectations or preferred procedures for emergency incidents and assumed that the crew would follow departmental SOPs. When the company arrives it is clear that this is a working fire. Because it is a daytime incident, the officer is relatively confident that the dwelling is unoccupied.

Upon arrival, before the officer can make a brief size-up, the crew goes into action. Two fire fighters pull the 150-foot, $1\frac{3}{4}$-inch hose line and extend it to the side door (fire can be seen coming out the windows near this door). Before the officer can join the two fire fighters extending the hose line, they have forced the door and entered the house.

When the officer catches up, one of the fire fighters has gone off on his own to search for any victims. The fire was confined to a room and its contents, and was extinguished using a few gallons of water. After the fire, the two fire fighters discuss with pride their ability to get in fast and put the fire out before anyone else can arrive on the scene.

Discussion Questions

1. Describe any potential disciplinary problems in reference to safety or SOPs that would be apparent in your department.
2. Discuss your own personal thoughts about the situation.

3. In a similar situation, how would you instruct your crew in their actions during emergency operations? Be specific.

References

Bruce, Neil. *Public Finance and the American Economy*. Reading, MA: Addison-Wesley, 1998.
Edwards, Steven. *Fire Service Personnel Management*. Upper Saddle River, NJ: Brady/Prentice Hall Health, 2000.

Fisher, Roger & William Ury. *Getting to Yes: Negotiating Agreement without Giving In*. London: Arrow Business Books, 1997.

Gray, John. *Men Are from Mars, Women Are from Venus*. New York: HarperCollins Publishers, 1992.

IAFC. Why Do Volunteers Volunteer? *IAFC On Scene*. July 13, 1995, pp. 1–3.

IAFF. *Politics and Legislation*. http://www.iaff.org/politics/us/content/bargaining.htm (accessed August 3, 2005), 2005.

Internal Revenue Service. *Individual Income Tax Returns*. Washington, DC: Statistics of Income Division, Unpublished Statistics, 2002.

Jury Verdict Research. *Employment Discrimination: Verdicts, Settlements and Statistical Analysis*. Unpublished, 1999.

NFPA. *Firefighter Fatalities Due to Sudden Cardiac Death, 1995–2004*. Quincy, MA: NFPA Fire Analysis and Research, 2005a.

NFPA. *Firefighter Fatalities in the U.S.—2004*. Quincy, MA: NFPA, Fire Analysis and Research, 2005b.

U.S. Court of Appeals for the Fourth Circuit, *Benshoff v. City of Virginia Beach*, 180 F3d 752, 1999.

U.S. Department of Labor, Police and Fire Fighters Under the Fair Labor Standards Act (FLSA). http://www.dol.gov/esa/regs/compliance/whd/whdfs8.htm (accessed September 20, 2005).

U.S. Supreme Court. *Sutton v. United Air Lines, Inc. (97-1943) 527 U.S. 471 130 F.3d 893, affirmed*, 1999.

U.S. Supreme Court. *Kimel et al. v. Florida Board of Regents, U.S. No. 98-791*, 2000.

Vlahos, Kelley Beaucar. *Race-Based Policies Raise Fraud Concerns*. http://www.foxnews.com/story/0,2933,95647,00.html (accessed August 3, 2005), 2003.

6

Customer Service

Knowledge Objectives

- Identify the potential needs of customers.
- Understand market failures and public safety monopolies.
- Examine customer surveys and demographic changes.
- Recognize the best practices to serve the customer.
- Comprehend the importance of the customer's distress and grief after a fire or other emergency that has resulted in life or property loss.
- Examine changes that can improve the quality of service to customers.

Overview

In the business community, customer service equates directly to survival and increased profits. Those companies that can identify what the customer wants, and then focus on these desires, will be able to increase sales and profit. Companies have gone out of business by failing to keep an eye on the customers and their needs.

Going out of business is unheard of for the typical fire and emergency services organization. There are some exceptions however, especially in the EMS business where there are a good number of private ambulance companies. What is generally true is that those organizations that practice quality service are those that provide the best service to the public and are able to secure adequate funding from elected officials.

What Is Private? What Is Public?

Who should provide customers (the public) with emergency services? Should fire and emergency services be privatized, or is this the rightful responsibility of public agencies? What is the role of government regulation of fire and emergency services providers? Should there be national standards for emergency services? Or is each local jurisdiction so unique that it should have its own tailor-made standards and customer service expectations?

Also, questions about the services governments can successfully and effectively provide for citizens need to be answered. Many Americans assume that the private capitalistic economy is the best place to find products and services. After all, this is what has made the United States great.

Others believe government should provide for many of the needs of its citizens, especially the more unfortunate. In general, most Americans prefer to rely on free markets, families, churches, and other private-sector associations whenever these can do the job. Governments are called on mainly when private-sector institutions prove to be severely inefficient and when they fail to satisfy certain minimum standards of fairness. Americans have built a strong economy based on competition and the free enterprise system. This is our heritage.

Justification for Government Intervention

When can government legitimately intercede to provide services, goods, or safety to promote its citizens' (customers') well-being? One type of justification occurs when there is a market failure, which provides validation for regulation by government. For example, if education for young children was provided exclusively by private *for-profit* institutions, the children whose parents could not afford the tuition would not get an education. In the United States, a public education up to and through high school is a public good that is provided free to the student and supported by taxes.

Another common example concerns traffic laws. If everyone were allowed to drive as fast as they wanted, then the public good of having safe roads for everyone would not be possible. Therefore, government regulations (such as speed limits) help maintain reasonably safe highways (**FIGURE 6-1**).

Fire departments have a monopoly for services they provide in the community. Because there is no direct competition, improvements, service levels, and fiscal efficiencies are under the control of the department and/or the municipality that oversees the department. In other words, if service is below nationally recognized professional standards, the customer has no competing department to compare to or select locally. Unless the public demands better service (remember, they think the majority of departments are doing a good job) or there is a regulatory body to oversee the operation, the fire and emergency services organization sets its own level of service.

Fortunately, most fire and emergency services organizations are dedicated to community service. However, the administration is often at a disadvantage when they would like to make a change that will bring improved service. In the normal business world, if the manufacturer does not make the best product at the best price for the customer, it will lose market share and profit or even go out of business.

Government regulation and elected officials help to provide the balance needed for this monopolistic public service. They provide the fire and emergency services administration with the independent oversight necessary to assure the public they are getting professional quality service.

In addition, for overall guidance, a survey of customers may provide some insight into their expectations. If their expectations are supported by professional standards or expert advice, the administration can look for ways to implement the suggestions. If the public's understanding is inaccurate, the agency can fashion an education program to correct this erroneous understanding of the department and its service to the community.

Public Preference Surveys

One useful technique to assess the needs of the customer and support budget requests is the public survey. These types of surveys can be used to recognize how the public perceives the department's present service, identify areas in which they would like improvement or expansion of service, and determine how much the public is willing to pay for improvements in service.

For example, budget requests for enhancements that result in additional personnel are always hard to gain approval for because personnel costs are always the most expensive items in the budget. In addition, there will be competition from other government agencies, which will also be attempting to gain approval for more employees.

FIGURE 6-1 Government Regulations Helping to Maintain Reasonably Safe Highways

After estimating the total costs for the proposal, including all personnel salary and benefit costs, the administrator would simply express the unit cost per person, per taxpayer, or per household for the community. Then, the survey question that could be asked would be: "Would you be willing to pay X dollars more per year to bring all engine companies up to the staffing levels recommended by national professional standards?" There are many ways to word this question. It would be a good idea to provide detailed justification for inclusion in the survey or as an addendum. Of primary importance would be gaining the approval of the chief administrative officer and elected officials before actually doing this type of survey.

A public survey would be one method of convincing the appointed and elected officials to buy into the proposal. This should be the ultimate test of whether the public would be willing to spend their tax dollars on the fire and emergency services budget request. This is a simple example of the survey process. Although the results of this survey will still not guarantee approval, the chief can argue that this is a fair indication of the public's (customer's) wishes.

Surveys are a great technique for fire and emergency services organizations, especially in light of the high evaluations of their service from previous public opinion surveys. Other agencies and their supporters may resist this technique and attempt to discount the results because any success may lead to budget cuts in their department.

Look for opportunities throughout the year to communicate with citizen groups, the press, and elected officials to gain support for existing programs or enhancement programs that the administration would like to propose in the future. When these opportunities occur, they're very short-lived. Be prepared with all the information for justification and an accurate estimate of all costs.

In most cases, the public, press, and officials have little knowledge or background when it comes to fire, EMS, and emergency preparedness services. In a survey commissioned by the National Fire Protection Association (NFPA, 1996), most people feel they are more at risk of dying in a fire in a hotel room than in their own homes. Actually, the opposite is true. This misconception may be the result of the fact that most home fire deaths do not get much media attention because they occur one or two at a time, as compared with incidents that have a high life loss such as an airplane crash. Over half of the respondents wrongly assumed they would have plenty of time to escape a typical living room fire. A majority had no idea how fast smoke, toxic gases, and heat can fill a room, blocking escape routes and reducing visibility to zero. For surveys to be meaningful, the persons completing them must have some idea of what constitutes quality service.

The first step may be to educate the public in two areas. The first is fire safety education, to teach them the dangers of a hostile fire in their homes. The second step, education on how to evaluate the public fire and emergency services, may be a little tricky because it tends to be complex. Giving the public the knowledge of how to assess the public fire and emergency services is necessary if they are to be able to have an informed opinion about the department and its services.

Community Demographic Changes

Monitor the community for demographic changes that may affect the services requested and tax revenues available. For example, older communities may see both a decrease in population and an increase in the number of senior citizens. Suburban areas are experiencing growth with strong support for improved services in areas such as education. Most growth areas will see an appreciable number of new families with school-aged children.

These demographic statistics are published yearly as estimates and can be found at state or local planning agencies. They are not 100% accurate on a yearly basis, but are very close in most cases.

Expanded Services

The types of services a department can offer affect its relationship with its customers. Many departments look at the addition of services outside of firefighting as a way to market the department to its customers. These organizations are quick to point out to citizens and elected officials that they provide extra value for the tax dollars spent.

The following is a list of services that a fire and emergency services organization can provide its customers. These are the primary areas to focus on when planning customer service efforts. Several of these may be the primary responsibility of other agencies as dictated by local ordinances or historical evolution, but all are provided in one fashion or another.

- Fire suppression
- Emergency medical service
- Technical rescue service
- Major disaster mitigation and recovery
- Prevention and education (fire, medical, and major disasters)

These will not be discussed in detail except to mention that there are standards for professional quality service for many of these customer needs. Following is a discussion of two examples of customer service.

Fire Suppression: Response Time

The capability of the fire department to respond with sufficient trained personnel and equipment in time to rescue any trapped occupants and confine the fire to the room of origin is a primary customer service goal. The term *in time* has been controversial. Until recently there was no consensus on its meaning in the American fire service; however, it is now defined in the NFPA 1710 standard.

Another measure that can be used by jurisdictions is to look at the existing response time and evaluate all areas of the jurisdiction to see if they are within the average response time for the jurisdiction. The department may want to benchmark its response times along with other local departments, and compare them against a national survey that reports response times. Although not very common, there have been some national studies in this area, such as the *Phoenix Fire Department National Survey on Fire Department Operations* (Phoenix Fire Department, 1999). The rationale behind this type of evaluation is that all citizens (customers) deserve equal fire protection as measured by response time. This type of analysis can point out areas that need new stations as well as areas that could be served by fewer stations and still meet the jurisdiction's average response time.

However, there are typically longer response times in rural areas that are sparsely populated. It becomes a cost-benefit calculation based on ability to pay. Areas with few residents will not be able to afford the taxes needed to support shorter response times. For example, in built up areas, the Insurance Service Office (ISO) Grading Schedule uses a response distance of 1.5 road miles that generally equates to about a 3-minute response time. In other prominently single-family dwelling developments, ISO uses a 5-mile distance as a maximum for adequate protection (ISO, 2003).

The following paragraph from the NFPA *Fire Protection Handbook* (1997) sums up background information regarding evaluation parameters for response times:

> In any locale, to be even minimally effective in controlling a fire, the initial responding apparatus should reach the emergency scene in time to prevent "flashover" (a very rapid spreading of the fire due to the heating of room contents and other combustibles). The time from ignition to flashover is controlled by many variables (e.g., the type of combustibles, the ventilation available, the source of ignition, etc.). Recorded times to flashover vary widely, but fire departments should operate on the assumption that prevention of flashover requires a response in less than 10 min. Where fire departments provide emergency medical service, the widely recommended 4-minute response for non-breathing or trauma victims is very important. That response time also is most helpful in fire situations. Given the unknown component of fire discovery and reporting its existence to the responsible fire suppression agency, recommending a "generic" response time is difficult. (NFPA, 1997, Chapter 10, pp. 2-6)

The NFPA 1710 standard, the ISO Grading Schedule, and the above paragraph can give some guidance to the planner or administrator trying to determine an appropriate response time for a jurisdiction. Remember that the location of fire and rescue stations is controlled solely by the response time chosen as a service indicator for the local jurisdiction. In general, longer response times will equate to fewer fire/rescue stations, and the opposite occurs when response times are lowered.

The total number of stations is very sensitive to the response time used. For example, let's assume that the existing response time (just travel time) is 4 minutes. If the required response time is increased by 1 minute to 5, there will be a need for 36% fewer stations to cover the same area. If the response time is lowered to 3 minutes, then 75% more stations will be needed. These are large variances.

For customers to be able to make a rational judgment about the quality of their emergency services, they should be able to understand the previous discussion on response times. It is very important that these citizens understand that uncontrolled fires grow exponentially, which is a mathematical term meaning that with every minute a fire grows faster than the previous minute. A fire can actually outrun its victim(s). The NFPA has several excellent videos that illustrate visually the rapid growth of fires. It would be a good idea to purchase these videos and use them in public fire safety education programs. There is no way of giving customers and elected officials the experience of an actual fire, but these videos come very close.

Fire Prevention

Fire prevention is a major customer service area for the public, the owners of buildings, and fire fighters (safety). The NFPA *Fire Protection Handbook* (1997, Chapter 10, p. 2) states:

> One basic aspect of a comprehensive public fire protection plan is the concept that it is infinitely better for a community to prevent fires altogether, or to mitigate them automatically through fire safety education and built-in fire protection features, than to depend solely on the fire suppression capabilities of the community's fire department. The goal of reducing the incidence and effects of fire, then, involves all aspects of fire prevention.

The customer has very little appreciation for fire prevention and therefore may not be as inclined to support funding of prevention programs. This conflict between funding emergency operations and funding prevention programs is very difficult to deal with in most fire and emergency services organizations. Advocates in the organization for funding and supporting emergency operations are always stronger than those supporting prevention programs. The beneficiaries of a comprehensive prevention program in many cases do not even know that they have been saved or property loss has been averted. This is customer service at its most efficient level, before a true emergency has started (**FIGURE 6-2**). However, in most cases the outcome is invisible.

FIGURE 6-2 Benjamin Franklin (1706-1790): "An Ounce of Prevention Is Worth a Pound of Cure"

The following article from *The Voice* (Drennan, 1998) points out the importance of fire prevention programs:

Forty More Are Alive Today

By Vina Drennan

Captain John J. Drennan died fighting fire in the city he loved with fire fighters he loved. For forty days his struggle to live was followed by the people of New York City who showed their respect for Captain Drennan and for all fire fighters who continue to put their lives on the line. Heroes, every one of them.

When the Fire Department of New York City (FDNY) chose to honor the commitment of John Drennan with a medal, I was appreciative of the honor. "Could we use the medal to honor a member who shows extraordinary commitment to reducing fire through prevention and safety education?" I naively asked. You can imagine the reaction and so the Captain John J. Drennan Medal is imprinted with the solid powerful word VALOR, and I sincerely am honored to present it to a fire fighter who risked much to be deserving of this medal. Yet as I stand on the podium at the FDNY Medal Ceremony and look into the faces of the fire fighter's family and see the pride in their eyes, I wonder if they even realize how close they come to losing their loved one every time a fire fighter responds to the alarm.

Yes it takes courage to fight fires, a courage I will never know, courage to face the flames, the heat, the darkness, the unknown. However, I am sorry I did not hold my own in those early days following his death because it takes great courage to fight fires with common sense too, and there are no medals for common sense in New York City. There are probably no medals for common sense in your department either.

You readers of this column know what I mean. You are continually overcoming the obstacles of the culture that sees suppression as the only way to fight fire. Statistics verify that in cities that commit a mere 3 percent of their fire budgets to fire safety, lives and property are saved. It's hard to tally the lives saved in a community where the commitment to install smoke detectors [and sprinklers] is a priority. We do not know the name of a child that is spared the scars of fire because one of you went into a school to teach the importance of developing an exit plan for their family. We do not know which elderly person you saved when you visited the senior citizen center in your neighborhood and gave a cooking safety demonstration. We do not know the names, but we do know the fatalities are decreasing and it is not because of bigger and shinier apparatus. Fatalities are going down because of the courage of fire fighters who stand up to the macho mentality that has driven our departments for too long. Fatalities are going down because there are fire fighters throughout our country who see a better way; members who truly want to prevent suffering and are tired of glorifying the failures of prevention. Many of you have witnessed too much in your careers, and you want it to stop.

Within FDNY there are also people that share the commitment to prevent fire. They quietly do their work; often without funding, without the respect which we honor the others. The results of their efforts are showing up. In 1995, more New Yorkers died as a result of fire than all the victims that perished in the bombing at Oklahoma City.

We lost 173 of our people that year but fortunately the trend has reversed as the department begins to focus attention on education and community outreach. At midyear 1998 we had 48 fatalities, way lower than the 96 deaths we counted as of June 30, 1996. Forty more people are alive this year and something has changed in New York City. Sure, we can give credit to the bunker gear fire fighters now wear that allow them to save more lives, and we can give credit to the use of defibrillators, which save more lives too. However, we must also give credit to the fire safety educators and the programs that are effectively preventing fire throughout the City.

I wish we could present the Captain John J. Drennan Medal to that Fire Marshal, who having spent much of his career extinguishing fire, now devotes his efforts to arson investigation and whose constant prayer is to have a month in which not a single life is lost. I wish we could present the Captain John J. Drennan Medal to the lieutenant who spends his day finding the materials he needs to take the fire awareness message to our people. I wish we could present the Captain John J. Drennan Medal to the fire fighters that distributed 23,000 batteries in a recent smoke detector program.

To whom would you give the Captain John J. Drennan Medal in your department? He or she is there somewhere, probably frustrated, maybe being teased or ignored; used to being treated as an annoyance. He's there. I wish I could come hang the Captain John J. Drennan Medal around his neck. "Don't Quit," I would say. "Please keep trying and don't lose your courage. Together we will make a difference."

Source: Reprinted from *The Voice,* Copyright© November-December 1998, The International Society of Fire Service Instructors. Reprinted with permission. All rights reserved.

Note: Mrs. Drennan was very happy with the 40 lives saved. However, in 1998 the fire deaths in New York City were down to 107, or 68 lives saved in 1 year. These numbers become even more impressive when you look at the number of fire deaths in New York City in the late 1960s, which were over 300 per year. There are a lot of customers who are alive today thanks to the efforts of modern fire suppression and prevention. This is the ultimate goal in customer service (FDNY, 2004).

Home Fire Sprinkler Systems

Home fire sprinkler systems may be the ultimate customer service for fire safety in the future. They are so effective, they are recommended for life and property protection from fire over smoke alarms. Although home fire sprinklers have been around for many years, they are installed in less than 1% of new homes. Part of the reason for the low installation rate is that home fire sprinklers are rarely required by building and fire codes. However, this has changed recently with the NFPA approving a provision in their *Life Safety Code* in 2005. Now we will wait to see which states, cities, and counties adopt this life-saving standard.

There has been steady organized national opposition to the code changes needed to mandate these safety devices in homes. When first proposed, the cost per square foot was expensive, but now it is common to have installation in new homes at $1 per square foot or less. Some, if not all, of the installation costs can be recovered by reductions in fire insurance.

A recent study of Scottsdale, Arizona's 15-year experience with residential home sprinklers (41,408 homes) showed the following results:

- "According to the National Fire Protection Association (NFPA) more than 80 percent of all fire deaths occur in homes." In Scottsdale, many lives have been saved in these homes protected by residential sprinklers.

- "The average fire loss in the homes protected with sprinklers was $2,166, compared to $45,019 in homes without sprinklers."
- "When you look at the Scottsdale data, 90% of the fires were contained with one sprinkler."
- ". . . the cost to install a fire sprinkler system is between 1 and 1.5 percent of the total building cost." (Home Fire Sprinkler Coalition, 2005)

Fire Safety Inspections

Many fire and emergency services departments use line personnel to also provide fire inspections of existing buildings. This is a good use of any stand-by time; however, time conflicts seem to be more and more common for station personnel. Beginning with fire calls then adding EMS responses, training, physical fitness, station and apparatus maintenance, and fire prevention duties can result in some critical nonemergency functions not being done or being done marginally.

Even though all of these objectives are very important and more than justified, in a busy fire company there will not be time for everything. Therefore, priorities must be selected. Number one has to be responses to emergency calls and the training (including physical fitness) that prepares these companies to do an efficient and effective job. Along with the skills training, prefire plans and familiarization inspections are a great help and can even save fire fighters' lives. All of these functions are necessary if the fire and emergency services are going to be able to give 100% effort when responding to a structural fire.

There is no better time to improve (or damage) the professional reputation of an agency than during fire inspections or prefire planning visits. Train and educate inspectors and fire fighters well and often in public relations.

The greatest chance for a bad impression will come if a correction order must be issued. Do it professionally. Make double sure that the violations are documented and validated. After the inspection, for those items that do not need immediate correction, go back to the office and research any potential code violations. Nothing is worse than referencing an incorrect code violation. It will destroy any respect and trust the public has for the department's program.

Be amiable but professional when presenting the findings of the inspection to the owner or occupant (**FIGURE 6-3**). A professional fire safety program will have a favorable effect on the ability of the fire department to obtain funding for needed equipment, personnel, salary raises, and new stations.

Fire fighters also gain knowledge that is helpful in firefighting. In conducting fire inspections and prefire planning surveys, the fire fighter has an opportunity to learn about conditions and construction features of buildings in the community before a fire occurs.

Enforcement of Fire Safety Codes

Enforcement is really a system of separate steps that add up to an effective program as follows:

1. *Code adoption:* Publicize the enactment of new or updated fire safety code ordinances. Highlight any potential controversial issues. Give a presentation outlining the major proposed changes and invite architects, engineers, developers, owners, and the media. Do not try to hide something in code language that may not become obvious until after adoption. The reaction will be strong and swift from developers.
2. *Plan review:* For new construction, it is imperative that the building be designed to comply with all fire safety codes and standards. Also, in the typical building code, about one half of all requirements are fire safety–based. Because the fire and emergency services agency has to respond to any fire emergency and put their fire fighters in harm's way, it has a greater motivation to make sure the building is constructed in accordance with the fire safety aspects of the

FIGURE 6-3 Presenting Inspection Results to Homeowners

building and fire codes. If the jurisdiction uses the NFPA *Life Safety Code*, it is a good practice to have the fire and emergency services enforce this code. In other jurisdictions that use only a building code, the fire and emergency services plans examiner can be tasked with reviewing fire hydrants, access to the building for fire apparatus, fire protection systems, means of egress, fire-resistive construction, and smoke control systems.

3. *Fire inspections:* The inspector should be tasked, again, with enforcing those items in the building code that are fire safety–related. If some code items are missed during construction, it may be impossible to gain compliance at a later date. The success of these inspections is directly related to the correctness and thoroughness of the plan review.

4. *Maintenance fire inspections:* These are needed to assure that fire protection systems, means of egress, and other fire safety items are properly maintained. To do this task properly, it is mandatory that inspectors know the fire code requirements and the fire safety aspects of the building code. In addition, housekeeping items that, for example, could mean the accumulation of excessive amounts of combustible trash in the structure must be enforced. These inspections can be combined with prefire plan visits.

Home Inspection Programs

Although some fire codes contain fire safety requirements for existing homes, because of political realities and expectations of privacy that most people have in their homes, most inspections are done on a voluntary basis. For example, NFPA 1, *Uniform Fire Code*, contains a requirement for existing homes to have a smoke alarm outside of sleeping areas and on each level of the house (these may be battery powered). For those jurisdictions that have adopted this fire code or another with a similar requirement, the enforcement is more encouragement than strict legal action.

Most home inspection programs are primarily devoted to fire safety education. It is extremely helpful if smoke alarms can be given and installed by fire inspectors or fire fighters during these inspections. Remember, the home is where most people (customers) die in fires.

One good idea is for the fire and emergency services agency to contact local civic organizations and request donations to cover the cost of a give-away

smoke alarm program. This makes a great public service announcement. If necessary, some can be bought with public funds. In simple management terms, this is leading by example.

Obviously, for serious violations of fire safety code, entry can be gained either voluntarily or by a search warrant. As an example, through interviewing several teenagers who were found with illegal fireworks, it is discovered that an individual is selling fireworks out of their home. With this information, the department's investigator can go to the state's attorney and ask for a search warrant issued by a judge. In situations such as this, police should be assisting in gaining access and enforcement.

There are two ways to schedule home inspections. One is to advertise the service by cable and regular TV, local papers, and mailings. The advertisement would ask the occupants to call a central number to schedule the inspection. The second, and this can be in addition to notifications, is to have fire companies go into the field and knock on doors for entry. Go in pairs. Having a partner will help protect against any misconduct accusations. More than two may look intimidating and the public will ask why so many members are needed to do a simple inspection. A four-person engine can split up into two separate teams. If a paramedic unit is also in the field, the two persons are the right crew size for these inspections. It is a good idea to place signs on the fire/rescue apparatus to the effect that the fire fighters are doing home safety inspections.

Timing is very important in these visits. Many more people are home in the evenings than during the day. Therefore, the best time to do home inspections is between 7 and 9 p.m. It is not a good idea to go at dinnertime or too late when people go to sleep. Be prepared to return on a different day or time. If they are not home, leave a brochure on the door.

There have been many success stories for fire and emergency services departments that do home inspections. This is a program that has a good probability of preventing fire deaths of our customers.

Preventive Medicine

Similar to the fire safety problem, emergency medical problems can be prevented in many cases. It is always preferable for a person to lose weight, quit smoking cigarettes, and start exercising to help prevent serious illnesses. One big difference in fire prevention is that buildings can be designed with built-in fire protection systems such as automatic sprinklers, which help to ensure a fire-safe building. Although fire safety education is used to help with the fire problem in existing and nonsprinklered structures, preventive medicine education and treatment are the method of choice to treat diseases and injuries.

The fire and emergency services organization can provide the customer with some preventive medical services such as high blood pressure and cholesterol screening. These types of customer services can be effective at identifying symptoms that may lead to serious illnesses and win many friends for the agency.

Customer Service at Emergencies: The Extra Mile

Many experienced fire protection experts estimate that each person will experience a hostile fire in their home about once in a lifetime. At these relatively rare occurrences, the department can make a lifelong friend. The same can be said for the more common but equally traumatic medical emergencies. If a death has occurred, these painful events become even more distressing for those close to the victim(s). This grief is intensified if the victim is a young child.

Both fire and medical emergencies can bring grief and hardship to our customers. Department members can be very helpful as a source of comfort and in locating resources to help the family.

Fire and emergency services personnel should be provided with information and techniques for the grief process. For example, careless comments such as "I understand exactly how you feel" or "Calm down; everything is all right" can be hurtful to the bereaved. Short seminar courses can provide members with the knowledge and awareness they need to provide compassionate support in distressing situations.

The following is a list of services that may be provided to the victims of fire or other emergencies:
- Explain all fire and emergency services operations.
- Separately talk to owners and occupants.
- Salvage furnishings, especially personal items.
- Provide temporary lodging.
- Provide meals and clothing.
- Provide transportation.
- Notify relatives and friends.
- Notify religious services, if appropriate.
- Contact the Red Cross and Salvation Army.
- Provide temporary pet services.
- Contact the insurance company.
- Contact the police for security patrol.
- Secure the building.
- Install smoke alarms, if needed.

CASE STUDY: Tragedy Strikes

At 2 a.m. on a cold winter night, your department is dispatched to an apartment fire. The caller notes that her apartment is on the third floor and she can smell a strong order of smoke and the fire alarm is ringing. Numerous additional calls are received.

When fire companies arrive, they find a large fire on the second and third floors of a four-story garden-style apartment building. The initial dispatch is three engines, two ladders, and a battalion chief. The first engine to arrive reports a working fire and requests the assignment be upgraded. This calls for the dispatch of two more engines, one ladder, a heavy rescue, a second battalion chief, the safety chief, and a deputy chief.

After the fire is extinguished, and thanks to the smoke alarms in the building, all the occupants escape without any deaths. However, several occupants are burned and require hospitalization. In addition, 10 apartment units are so heavily damaged by fire that the occupants are not allowed to return. These people are now homeless.

Discussion Questions

1. Make a list of those services that the fire and emergency services personnel could either provide or direct these victims to for help, and identify a contact person and phone number (24-hour access) for each.

2. If an elected officials or the press show up, describe how they should be treated and what information should be gathered for their knowledge.

References

Drennan, Vina. Forty More Are Alive Today. *The Voice*, International Society of Fire Service Instructors, November-December 1998.

FDNY. *Civilian Fire Fatalities*. http://nyc.gov/html/fdny/html/stats/graph_fire_fatals.shtml. (accessed July 24, 2005), 2004.

Garvey, Gerald. *Public Administration: The Profession and the Practice*. New York: St. Martin's Press, 1996.

Home Fire Sprinkler Coalition. http://www.homefiresprinkler.org/hfsc.html. (accessed July 24, 2005), 2005.

ISO. *Fire Suppression Rating Schedule*. Copyright ISO Properties, Inc. Jersey City, NJ, 2003.

NFPA. *NFPA Fire Protection Handbook*, 18th ed. Quincy, MA: NFPA, 1997.

NFPA. Americans Underestimate Fire Risk. *NFPA Fire Journal*, May/June 1996, p. 77.

Phoenix Fire Department. *Phoenix Fire Department National survey on Fire Department Operations*, 1999.

7

Training and Education

Knowledge Objectives

- Identify training needs for fire and emergency services personnel.
- Understand the necessity for continuous reinforcement of basic emergency services skills and knowledge.
- Examine both local and national developments that require new training and education efforts.
- Recognize the best training practices that lead to safe, professional emergency operations.
- Comprehend the importance of training and education at both the fire fighter and fire officer level.
- Examine changes that can improve the quality of training.

Overview

Administrators are responsible for the efficient and effective emergency operations of the department. They should be aware that there are no shortcuts to having members who are well-trained, along with properly maintained equipment and standard operating procedures (SOPs) for emergency operations. Just like a great football team, success at the emergency incident is the result of a team effort in terms of training, equipment, and a plan. Now comes the important question: What does well-trained mean? Generally, it means that the providers have the ability to complete manual procedures with little or no conscious effort. This level of ability is achieved only after practicing the skill(s) many times, beginning with an intensive recruit school.

Athletes may always remember how to do the sports they excel at, but their ability will steadily de-

cline once they stop practicing. Even after many years of not riding, a cyclist will not lose the ability to ride a bike; however, the ability level will not be at its peak. Athletes perform their best with regular practice, increasing their dexterity along with physical strength and stamina.

To be proficient at emergency operations, members have to be thoroughly trained before participating, and must receive updated training regularly. In many fire and emergency services departments, some essential skills are used infrequently. A downward trend in the number of structural fires has been experienced in many fire departments. Some officials make the argument that because there are fewer fires, there is less value to maintaining minimum staffing and training for firefighting. Actually, the opposite is true. This profession requires a higher level of training because of the emergency nature of the service.

Workers in nonemergency jobs, such as auto repair technicians, can open up a service manual and search for the information they need to successfully do the job when faced with a challenging repair. The technician may even take some time to contact another more senior and experienced technician for advice. In emergency operations it is unacceptable to not do a job right the first time or to take too long to finish the operation, because it results in the potential for increased property loss and more deaths and injuries.

Emergency operation skills become even more critical when many of the tasks must be completed with a team of people. Training in the basic skills must be an ongoing and never-ending process. One specific area that always seems to be a problem in smaller paid and volunteer departments is finding the time and opportunity for live fire training exercises. To retain skills and safety, however, live fire training is essential.

As a minimum, the fire and emergency services organization should schedule drills for incumbents that cover all the basic essential job functions throughout each year. For example, National Fire Protection Association (NFPA) 1001 can be used for job performance requirements, which will cover the fundamentals of basic firefighting.

The less frequently a skill is used, the greater the need for practicing the skill. Also, it is common for members to have a better memory for the skill than their ability to actually perform the skill. Skills that require muscle coordination can degrade rather quickly. This varies for each individual, but because fire fighters work in teams, training must be directed to the least able member. For example, some fire fighters find it necessary to retie all the knots frequently to make sure they maintain their competency.

The New York City Fire Department is the largest and busiest fire department in the United States. In 1995 it instituted a Back-to-Basics training program designed to ensure that fire fighters maintain their skill level in the most important aspects of fire suppression operations while they receive training in new areas (**FIGURE 7-1**). Even with a large number of experienced fire fighters (as a result of the high number of working structural fires), this department saw a need to re-enforce basic skills by using a structured mandatory in-service training program (USFA, 1997, p. 13). If the largest, busiest, and most experienced fire department in the country sees a need for in-service basic training, it seems logical that all fire departments need this same type of training.

Actually, the fewer working fires that a fire department experiences, the greater the need for train-

FIGURE 7-1 Fire Fighters Stretching a Hose Line

ing in emergency skills and operations. All members should receive this training including those assigned to staff positions. It is not uncommon to have fire fighters working in staff functions, such as fire inspectors, who may not experience any firefighting for years at a time.

Traditional in-service training programs rely on company officers and their knowledge, experience, and individual approach to training. Even with the most competent officers, these programs lack consistency because of the numerous fire station locations and different shifts. The training given in one station on a particular day will not be the same as any other station. Therefore, any in-service training program should start with department-wide planning and implementation.

In-service training for the safety and efficiency of fire fighters should be a first priority. NFPA 1500, *Standard on Fire Department Occupational Safety and Health Program,* can be used as a basis for a departmental in-service training program. This should be in addition to basic skills training. Nationally derived standards are commonly used as a benchmark for fire service training and safety programs. The key to providing the best fire and rescue service is a comprehensive, exhaustive recruit school for new fire fighters and an in-service training schedule that stresses and reinforces basic skills.

The NFPA 1500 standard calls for monthly structural firefighting training totaling a minimum of 24 hours per year. This is a minimum number of hours and would be appropriate only in departments that protect a small number of single-family dwellings. Most of these drills can be given at the fire station, ensuring a convenient location without requiring the company to go to a training facility. There are times when fire companies must go to a central training facility, such as for live fire training, and the department may have to make provisions for a stand-by crew at the station(s). For example, one metro department provides a special shift of four 10-hour days that fills in for companies that need to attend multi-company drills or paramedic recertification classes.

Some larger departments use a cable TV system to distribute training videos and lectures to their many stations. It is not unusual for a station to have to respond to an emergency call during the scheduled broadcast time, causing missed training segments during their absence. For this type of training program, the department should purchase a VCR and TV for each station that will be dedicated to training. The VCR would tape each training program as it is being broadcast. For those stations without cable TV, a tape could be prepared and delivered to the station so the fire fighters could watch at any convenient time. Also, there are a growing number of commercially prepared training video packages available from many sources. Many departments purchase one or more copies for a library and distribute the tape to each station on a rotating basis.

Some recertification training for EMTs and paramedics could be delivered by the cable TV system to on-duty personnel and taped for viewing by other shifts (especially lecture presentations). Using telephone (or video) conferencing, these lectures could be interactive with the students many miles apart. Practical and written exams may still have to be administered in person or supervised tests can be taken over the Internet. This technology has the greatest potential for reducing the off-duty time needed for recertifications. As with any new technology, changing the institutionalized certifying agencies will be difficult, but not impossible.

Another new technology in delivery is the use of interactive multimedia training. At the present time, there are only a few fire and emergency services training programs being delivered by CDs or interactive Internet programs. These are especially useful for required yearly recertification training such as hazardous materials. The big advantage is the ability to obtain the certified training at the local fire station or at a member's own home. Some of these computer-based training packages even have the ability to give a test and record the results in a tamper-proof computer file. The disadvantage is the initial cost of multimedia computers, and the training packages can be pricey. However, if costs are compared to classes that require a certified instructor to provide the same training, the interactive program would be a winner in both cost and convenience.

For station drills, one effective technique is to have the new rookie give basic drills. This new fire fighter will have recently learned the most current methods at the training academy. In many cases, training agencies are the first to adopt new skills and equipment. This also helps reinforce the skills in the new members.

For EMS skills, there is a national standard for both training and testing. Formal training and certification is needed to become an EMT or paramedic in all states. Training requirements are prescribed in a national curriculum created by the U.S. Department of Transportation, National Highway Traffic Safety Administration. For example, the Emergency Medical Technician-Basic has a prescribed national standard curriculum along with a minimum number of hours needed for the training (110 hours). It does not make a difference if an EMT-B is in a city, county, or state agency, or if the area served is urban or rural; each person will have essentially the same training and certification testing.

In the emergency management profession, the Emergency Management Institute of the U.S. Federal Emergency Management Agency provides a nationwide training program of resident courses, interactive Internet courses, and nonresident courses to enhance U.S. emergency management practices. There is also the national Certified Emergency Manager program, which requires the applicant to demonstrate formal education, training, practical experience, professional contribution to emergency management, examination, and participation in full-scale disaster exercises. The International Association of Emergency Managers sponsors this certification.

Knowledge and Cognitive Skills

Technical competencies are most successfully taught to the adult learner through hands-on training. Many higher-level skills such as emergency command, hazardous materials evaluation and operations, fire prevention, and emergency medical service can be best learned by the adult student through acting out scenarios. The highest level of knowledge retention will come as a result of actually practicing a skill or administrative task, even if only a made-up scenario is used. Although many of these skills require massive amounts of special knowledge, the best retention comes after applying the knowledge to an actual incident or training exercise. This is especially true with the adult learner. This type of learning exercise also tends to keep the interest of the adult learner better than just memorization of facts and regurgitation on an exam. Some multimedia presentations and reading assignments are necessary in the familiarization process, but the actual exercise is more important for understanding and long-term retention.

There are many training packages available either as computer software or in other distance learning formats. These are very helpful, but an exercise should be included in the program of instruction to assure long-term retention and immediate recall at the emergency incident.

Higher Education

Typically, the majority of higher education programs are offered at community colleges, with a few universities offering bachelor's degrees. Many community colleges have a committee made up of representatives of the fire and emergency services organizations, of which the students are members, to provide curriculum advice. The National Fire Academy's Higher Education program has developed a recommended core curriculum for fire science programs that would be good to follow. The model curriculum can be found at the U.S. Fire Administration Web site (http://www.usfa.fema.gov/training/nfa/higher_ed/).

When shopping for a college-level education, it is very important to make sure the program meets your needs. You should do a quick check to see if the college is accredited by contacting the U.S. Department of Education or Council for Higher Education Accreditation. Each has a list of colleges that are accredited. Institutions that are recognized by a *regional accrediting agency* are accepted at almost all colleges and universities. Other accrediting agencies and individual courses that are recommended by The American Council on Education (ACE) are *not* accepted at many of the upper tier schools. Check carefully that your course work will transfer if you are planning to go to another institution of higher learning.

In addition, look at the courses. Many offer only fire science and do not have courses on EMS or emergency preparedness management and operations that may be of interest to anyone who would like a well-rounded education in all emergency services (except law enforcement or military). And finally, look at the official degree title. Some add a small number of courses to an existing program, such as business or public management. A degree with an official title of Bachelor of Science in Fire and Emergency Services will mean more than a Bachelor of Science in Public Administration with a "concentration in fire science" simply listed on the transcript.

There are many forms of incentives to encourage members to take advantage of formal college-level education. Educational incentive pay and either credit or minimum educational requirements for promotional opportunities are good policies and are used by some departments.

Bachelor's degrees and some associate's degrees can be achieved via distance learning over the Internet (**FIGURE 7-2**). This is the most common method of instruction for fire and emergency services education at the bachelor's degree level because most students are spread out throughout the country.

There are some drawbacks to Internet-based programs. They require a great deal of self-discipline in setting aside the time for studying and completing the homework. Many of the younger members of fire and emergency services organizations can have trouble with these types of self-study educational programs because they have experienced a heavy use of video in both entertainment and education without much need for reading. These multimedia methods are also very time consuming and expensive to create. We are still a long way from extensive use of video or other multimedia methods at the college or university level.

Professional Development

Many organizations send their members to seminars or training sessions. This is a great method of increasing knowledge and gaining valuable experience from national experts on advanced or specialized topics. Again, because we have more than 30,000 fire departments in this country, it is impossible for any one organization to have all the knowledge and experience needed for every potential emergency situation.

FIGURE 7-2 Distance Learning Student

These training sessions are sometimes provided at the local training academy by outside experts. The local fire department may be fortunate enough to have a nationally recognized expert and should make use of this resource. The training bureau could even videotape the session for those who cannot physically attend, for circulation, or for future use.

There are numerous state, federal, and national training programs and conferences that have excellent content. When the department sends members to these educational programs, have them agree to teach their new knowledge to all members when they return home.

Training and professional development is a never-ending process. It starts at rookie school and continues throughout the member's career.

Recertification

One weakness in most certification systems and the NFPA professional standards is that there are no recertification or continuing education requirements. In the future, there may be requirements for existing members to update their knowledge and skills on a yearly basis. For example, the State of Florida requires people holding the fire officer, fire inspector, and fire instructor certifications to complete 40 hours of continuing education every 3 years.

In areas where there is no requirement for recertification, the administration can include items in the yearly performance appraisals that will require self-study or attending seminars and formal courses to update knowledge and skills. Over a 20–30 year career, many things change and it is very difficult to retain information that is used infrequently.

Each department should create a standard training program and schedule. This approach will help to assure that all members have equal abilities, knowledge, and skills and therefore can be easily moved from assignment to assignment. This will also ensure that an engine company that services a rural farm district, when dispatched or transferred to an area with high-rise buildings, will be able to operate with efficiency and skill at a major fire.

Standard Operating Procedures

Standard operating procedures (SOPs) are very common in fire and emergency services organizations. There is a reliance on SOPs because of the nature of emergency services. In many cases, split-second decisions must be made by the first arriving company officer that can have a huge effect on the successful outcome of the incident. People can die (including fire fighters) and greater property loss will result if the proper actions are not completed in the first few minutes of emergency operations.

Once SOPs are adopted, training sessions must require that all members are proficient in following the procedures. Any actions or situations that may require variance need to be addressed and incorporated into the SOPs and training sessions. SOPs also need to be a major reference for questions on any promotional exams to assure that all officers are familiar with all the requirements.

To illustrate the complexity of initial actions, the following is a list of major questions that could confront the officer of a first arriving engine company at a single-family dwelling fire:

- Should a supply line be laid for water or should tank water on the engine be used?
- From the first observations, can it be determined where the fire may be in the structure and if there are any potential victims inside?
- What size hose line should be used?
- What type of nozzle would be appropriate?
- Are there enough fire fighters to attack the fire safely?
- If the building is locked, how do the fire fighters gain access?
- Is there any information or observations that need to be communicated via radio to other responding companies or the incident commander?
- Which door should be used to enter the structure for fire attack?
- Does the building show any indications of a backdraft or flashover condition?
- Are there any signs of any structural weaknesses?
- What method will be used to account for the progress and location of all fire fighters?
- Can it be determined if more resources will be needed than are presently either on the scene or responding?

These can be complex questions to answer, with numerous variables, and many of the variables are typically unknown when the first fire company arrives. Many fire and emergency services organizations have an SOP that covers the typical situations and allows the officers to vary the SOP if the situation dictates. For example, most departments require a minimum size hose line to be used when attacking a structural fire. Even though many fires can be extinguished with a very small amount of water, this is not known until after a few minutes on the scene for an investigation when the company has already stretched its initial handline. Therefore, if fire fighters stretch a handline that is too small, the fire will intensify and become out of control before a second, larger handline can be extended.

These SOPs should be updated or reviewed frequently for both their technical correctness and their relevance. For example, when a first version of an SOP on handlines for structural fires was created, it might have specified the minimum size line as a $1\frac{1}{2}$-inch line. After doing some research on newer hose sizes and nozzles, the department may have decided to equip each engine with a 2-inch line capable of flowing 250 gallons per minute. In actual testing using the department's personnel, it has been shown that a two-person crew can extend and operate this handline (using a solid stream nozzle) with a little extra effort. The research has also noted that even though this larger line is available for fires that are beyond the capability of a $1\frac{1}{2}$-inch handline, the crews typically stretch the smaller line out of habit. Therefore, to eliminate the possibility of an incorrect (5-second) decision, the fireground SOP is changed to require the use of the 2-inch line on all nonresidential structural fires.

This same type of analysis was used by most fire departments when adopting SOPs that eliminated the use of "booster" ($3/4$- or 1-inch) hose at structural fires. The booster hose was easy to pull off the engine but was found to be lacking when used on challenging fires.

Consistency and Reliability

SOPs are the basis for consistent and reliable quality and levels of service. They should be followed with few exceptions, and must be studied, reviewed, and practiced frequently. "If the bureaucracy is to operate successfully, it must attain a high degree of reliability of behavior, an unusual degree of conformity with prescribed patterns of action. Hence, the fundamental importance of discipline, which may be as highly developed in a religious or economic bureaucracy as in the army" (Garvey, 1996, p. 132).

Engineers achieve consistency and safety if their designs follow scientific principles and include a safety factor. There is also a margin of safety that is provided in the conformance to SOPs. SOPs are rules that clarify consistent actions that result in proven outcomes and safe practices.

Developing SOPs

There have been numerous arguments and discussions over the years that revolve around different fire tactics and equipment. For example, the type of nozzle to be used (solid stream or fog) is one area of disagreement.

For controversial subjects, the chief should create an advisory committee that can analyze and make recommendations on SOPs. In addition, contact other fire and emergency services agencies and request copies of their SOPs. Do a little background research and find out if the department's SOPs are actually followed or if there are enforcement issues.

This committee would be instructed to find the best solution with the understanding that exceptions may be handled in the field. For example, after listing the pros and cons for a particular policy, the committee may find out there is no *right or wrong* solution, but a lot of gray area. The committee, with the final endorsement of the chief, would then select the policy that worked for the majority of the cases. To expect to find a solution that will work 100% of the time is unrealistic. With well-trained and educated fire officers, variances in the field, although they should be rare, can be expected and encouraged when justified.

For example, a fireground SOP may detail that the first due engine must take actions to extinguish the fire. On one rare occasion, however, when approaching an apartment building, an officer of the first due engine spots a woman holding a baby at a third-story window. The officer would order the crew to remove the ground ladder from the engine and use it to rescue the victims. The officer should immediately radio to other companies and any responding chief officers that his or her company had varied from the SOP and another company must accomplish the original assignment. Any time a variance occurs, it needs to be communicated immediately to the command officer and other companies that are responding to the same incident.

Regional Approaches to Training

Fire and emergency services in the United States are often provided by smaller departments (departments with fewer than six fire stations). Often the local department does not have the resources, such as instructors, instructional materials, or time, to properly train its members. This is a structural characteristic of the U.S. form of government, which delegates many services to the local government level. Remember, there are over 30,000 fire departments in the United States.

Larger departments can also need training from outside sources. This is a good indication that the department is progressive and knows when it needs support from other training sources because of lack of experience. Areas of specialized skills, such as hazardous materials, terrorism, technical rescue, wildfires, catastrophic disasters, high-rise fires, multi-alarm mutual aid incidents, and large-scale emergency medical incidents, are all infrequent events in most jurisdictions. Anyone can read a book about someone else's experiences, but a person with actual experience can really help the training program and the retention of information by the students.

No fire and emergency services agency can handle all its emergencies itself; it will need help sometimes. Even the largest departments must call on their neighbors for help when a major emergency incident occurs. By mixing companies up in training exercises, exchanging instructors, and promoting other regional efforts to share training, the departments will more easily operate together when a large-scale emergency calls for them to work together.

As mentioned previously, many fire fighters are not given the needed training at the basic skills level. An NFPA study indicates that an estimated 233,000 fire fighters lack formal training (NFPA, 2002). Much of the needed skills and proficiencies training is not available at the local level. In many parts of the country outside of major metropolitan areas, training is made available by county or state fire training agencies. However, most of this training is voluntary.

The profession of firefighting is still evolving. The first national standards did not evolve until the early 1970s. Acceptance has been slow, but steady. One example of the difficulties in gaining acceptance can be seen in the different hours of fire-fighter training needed to achieve NFPA Fire Fighter I/II certification, which range from 160 hours to over 560 hours in different states, counties, and cities. NFPA standards do not contain any recommended number of hours for a student to become competent. Because fire fighters do not become proficient without repetition, the lower number of hours for the basic firefighting courses should be suspect as not being effective.

Training to Fit a Need: An Example

Hazardous materials emergencies have required a rethinking of many of the traditional approaches the fire service has relied upon in firefighting and rescue work.

The speed with which an incident is approached, the level of pre-incident planning required, and the dependence on specialists from other fields including the private sector are all radically different from the SOPs required in firefighting. This rethinking extends into the heart of hazardous materials team planning and justification and leads us to ask: Does every community need a hazardous materials response team of its own?

In the past two decades, there has been an explosion in the number of hazardous materials teams created in the public sector. In most cases, these teams are under the jurisdiction and command of a local fire department. Yet, since these teams have been formed, many communities have found that these specialized teams are used rarely, if at all. For example, the Houston (Texas) Fire Department is recognized as one of the busiest departments in the United States, and because of the high density of oil and chemical industries, they are one of the busiest haz mat units in the world. In Houston, this amounts to approximately 956 runs in 2004. In smaller municipalities, a single call per year is not unusual (*Firehouse,* 2005).

According to incident records collected from state and local jurisdictions over the past years and discussions with members of various operational teams, hazardous materials releases account for 1% of the fire calls made by a local jurisdiction. Approximately 50% of the incidents involve very small quantities (less than 55 gallons) of the hazardous materials, and 75% of these incidents occur at fixed facilities where the material is routinely stored and used.

In the face of few calls, limited chances to test their training, and an increased need to broaden their experience and knowledge in the field, hazardous materials team members become discouraged and burned out. This increase in demand and costs for training often reduces the team head count to less than optimum levels. Department chiefs with few hazardous materials–related runs have difficulty justifying the continuing cost of training and investment in new technologies and equipment that keep a team at the highest state of readiness for any potential emergency.

In the initial rush to create hazardous materials teams everywhere, it was not possible to accurately evaluate the basic premise that assumed that every community needs its own team. Many experts originally supported the creation of many teams.

As with any municipal service, there are community criteria that mandate the operation of specific municipal services like the fire and police departments. However, there are many smaller communities who have no police and/or fire department of their own; this service is provided by a county or state organization or as part of a regional service alliance.

Cost is one of the primary reasons that many communities have consolidated their teams into regional or district hazardous materials response units. Given that a hazardous materials response team should consist of approximately 20 members to provide coverage on a 24-hour basis, including over holidays, weekends, and while some team members are in training, staffing costs, as well as the cost of training and equipment for each team member, are a major expense. A great example of a regionalized approach (several counties in each region) was created in Massachusetts and a map of their statewide organization is shown in **FIGURE 7-3**.

Other Ongoing Considerations for Hazardous Materials Teams

The kinds of training opportunities available in a community will also affect cost, as well as team member motivation and commitment. The availability of hands-on drills and field training with team members and those of any teams likely to be called in on mutual aid is impor-

tant. However, the organization of drills and training sessions requires a lot of coordination and networking within the community and with the public and private sectors. This time must be considered as part of the work hours of the team leadership and command structure.

The experience of a team is the most valuable resource available to it. If a team has little field experience, their fieldwork is shaped only by the controlled conditions of drills and training sessions. Despite the best intentions, these efforts still fall short of the hands-on learning of actual incidents. In communities with limited run numbers related to hazardous materials, local or regional mutual aid partnerships can be an effective way to provide appropriate levels of protection while reducing individual community costs.

This type of analysis can be used for any special service that a fire and emergency services agency considers offering or presently provides. Many special services can be better provided through the cooperation of several departments to share the costs and staffing. This can include services such as ladder or rescue companies that are difficult to staff and whose personnel are difficult to train adequately in a small department with few calls.

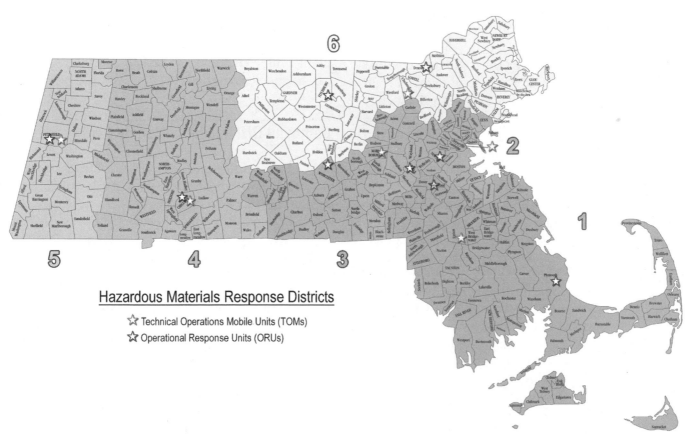

Hazardous Materials Response Districts

☆ Technical Operations Mobile Units (TOMs)
☆ Operational Response Units (ORUs)

FIGURE 7-3 HazMat Districts Map (http://www.mass.gov/dfs/er/hazmat/ distmap.htm)

Initial Fire Attack: Training Goals

This section provides an example of one portion of the successful extinguishment of a structural fire. This is a good example of how to determine acceptable competencies for one very important part of firefighting. Similar approaches can be derived for other emergency services operations. Many other skills and knowledge are needed to manage the many and varied emergencies a department is called to handle.

To determine the department's capabilities for initial fire attack performance at structural fires, NFPA 1410, *Standard on Training for Initial Emergency Scene Operations*, describes a method for evaluating the adequacy of the department's training program. This standard outlines typical fireground evolutions and recommended completion times. The fire company or department can evaluate the effectiveness of its training program by comparing the times that its crews attain doing the same evolutions.

These initial fire attack evolutions are broken down into three separate items: Handlines, Master Streams, and Automatic Sprinklers. In many cases, the initial actions of the first arriving fire companies will determine if a good stop can be made on a fire.

If the times that crews achieve are slower than those that are recommended in the standard, then the administration may want to evaluate three separate areas that have an impact on the proficiency of fire crews. The first would be the frequency of either training sessions or actual fireground experience. If fire fighters perform these basic evolutions frequently at actual fires, their skill level will be higher. If the fire experience is low, then the frequency of in-service training needs to be increased. The second is SOPs, and their critical importance has been discussed previously.

Finally, staffing can affect the times of each evolution. Staffing is described in NFPA 1410 as ". . . the average number of personnel that ordinarily respond" (NFPA, 2000a, p. 4-1.2). However, the staffing provisions listed in NFPA 1500, *Standard on Fire Department Occupational Safety and Health Program,* must be followed to assure the safety of fire fighters participating in the evolutions and on the fireground. For handline evolutions, NFPA 1410 explains, "A minimum of two fire fighters shall be used on each hose line to keep interior attack lines under control" (NFPA, 2000a, p. 2-2.4). This is similar to the wording in NFPA 1500, which states in part; "Members operating in hazardous areas at emergency incidents shall operate in teams of two or more" (NFPA, 2000b, p. 6-4.3). All three of these areas (training frequency, SOPs, and staffing) should be considered individually and in combination when analyzing any problems.

Another advantage of practicing these evolutions is that this will fulfill one of the requirements of NFPA 1500. The standard states, "Where the fire department is responsible for structural fire fighting operations, the fire department shall provide structural fire fighting training at least monthly" (NFPA, 2000b, p. 3-4.2). However, these evolutions should not comprise all of the monthly training drills, as they are only one essential part of a fire attack. Successful firefighting efforts involve many other skills and knowledge and must be reviewed periodically so that they are performed efficiently on the fireground.

Experience has shown that most fire departments can use their existing SOPs and fire apparatus to complete these evolutions within the suggested times with appropriate training. As noted above, some departments may want to install additional preconnected large capacity handlines and certain SOPs may need to be adjusted. The evolutions are described in performance objectives along with the appropriate maximum times. The maximum time benchmarks are reasonable enough that any fire department can achieve a passing mark with the appropriate training, SOPs and staffing.

Other goals can be formulated from standards such as NFPA 1001, *Standard for Fire Fighter Professional Qualifications*. This standard specifies the minimum competencies for a fire fighter. Other subjects can be added as needed.

Company Officer Education and Training

This level of officer is either a lieutenant or captain. Also, a chief in a one- or two-station department may have to operate as a company officer on many occasions. Because company officers must operate as part of the crew in many situations, training and education are both needed.

The following is a checklist, although not all-encompassing, of items that are helpful in preparing to function as a company officer professionally:

- NFPA professional qualification certifications
- Associate's degrees in fire science
- Promotional exams
- Officer candidate schools
- In-service training
- National Fire Academy courses
- Fire and emergency services textbooks (self-study)

- Professional periodicals
- Conference and seminar attendance
- Experience and common sense
- Visit other departments of all sizes and types; there are always good ideas everywhere.

Chief Officer Education

Chief officers need to build on their extensive experience and training provided during recruit, in-service training, company officer training and education, and formal higher education (bachelor's level is recommended) to be effective (**FIGURE 7-4**). As a prerequisite, the general educational (i.e., reading, writing, math, history, humanities, etc.) knowledge for the company officer is relevant to the chief officer and is offered at many community colleges. A chief officer should look at formal education as another tool to help be effective at their job. Many of the other agency heads, appointed officials, and elected representatives will have at least a bachelor's degree. To be on an equal footing professionally, the chief needs an equivalent education.

Several fire and emergency services bachelor's degree programs are available at universities and colleges. For example, Internet-based bachelor's degrees are available through the University of Florida—Fire and Emergency Services and the National Fire Academy (NFA)-sponsored Degrees at a Distance Program. A few colleges offer a traditional classroom setting for fire and emergency services, such as Arizona State University, California State University—LA, Colorado State University, and John Jay College. In addition, several universities offer bachelor's and graduate-level degrees in fire protection engineering.

FIGURE 7-4 Fire Administration Class at the Florida State Fire College

Conferences and seminars have both programs for basic skills and sessions for management and leadership education. These are great opportunities to be exposed to contemporary knowledge and education and to re-enforce basic management and leadership skills. One word of caution: There is no peer review of the content of these education and training sessions. Therefore, it would be good to confirm what is heard with others' knowledge and your own experience. Also, have an open mind and be prepared to accept new ideas that may be different than what was previously accepted as the truth. Many administrative procedures and techniques evolve and change over time.

Another great opportunity for an administrator is to be one of the presenters. With the appropriate knowledge and experience, the seasoned administrator has a lot of wisdom that can be passed on to others. Ideas become clearer as they are justified and explained to an audience. This is also a good time to research and review references that support the presentation. During the presentation, there may be questions or feedback from the audience. A good lecturer will try to learn something from every audience.

Quality Service

Training and education are crucial for quality service. Sometimes they are overlooked because of the nature of the people who perform emergency operations. We are dealing with people who were well trained when they first became fire fighters and may not be aware that their physical skills have waned over the years. It takes a strong commitment to train at the level needed to keep all basic skills sharp over 20 to 30 years of service. It also takes a disciplined and rigorous promotional and in-service training program to keep officers at their peak of effectiveness. As always, people problems are more difficult to deal with than equipment problems. It seems it is more common to replace an inefficient fire pumper than it is to train a seasoned fire fighter to the skill level obtained earlier in their career. Just as water pumps are tested once each year, testing fire fighters once each year is a very good practice.

Finally, higher education is the key to the door of opportunity both personally and as a chief officer. It puts the chief and any staff on equal ground with contemporaries, and that gives the chief many opportunities, such as to acquire a fair share of the tax revenues. In most cases, education will give the chief the tools needed to be a change agent.

CASE STUDY: Would Training Have Made a Difference in This Tragedy?

The following article (in part) was written after the death of a fire fighter and concerns the debatable circumstances of his death. This article opens the door to a very sensitive subject. Without facing uncomfortable situations, a person or organization cannot make needed changes.

Local "Squeaking": The Responsibility of Firefighter Training

By Billy Goldfeder, EFO

"If you're gonna go to fires, you must have relevant and regular training. Or, make the scene even safer and just stay home."

Lancaster (MA) Firefighter Martin McNamara was killed fighting a house fire at the end of last year. The issue of training continues following his death. While inappropriate for discussion the day after the fire, predictably the issue remains a hot topic now. The issue appears to be one not just specific to his tragic death but to all firefighters. His death has raised the issue to levels of visibility.

One news story about the McNamara tragedy said that the Lancaster Fire Department has an "informal and poorly organized training program." That may be true. But it isn't a whole lot different from many, many fire departments around the country. While so many areas are focused on other important training issues such as EMS, WMD, terrorism, haz mat, and numerous other related services, the BASICS OF STRUCTURAL FIREFIGHTING HAVE GOTTEN LOST IN THE SHUFFLE. What? A department with an "informal and poor" training program? A shock? As some say: PUH-LEASE!

Maybe Firefighter McNamara's department training wasn't as good as it could have been. That's not for us to say as of yet. But if it wasn't, it is no different from many departments—volunteer, call, or career. So maybe his tragic death serves as a wake-up call. Or, just like The Station Nightclub fire in Rhode Island—Rhode Island takes LOCAL action to change its fire codes while the rest of the states wait for "their fire" to give them their "local" wake-up call.

Discussion Questions

1. Do you agree or disagree with the author's statement "the basics of structural firefighting have gotten lost in the shuffle"? Support your answer with specific examples.
2. Would you support mandatory fire-fighter training including on-the-job refresher training by either the state or federal government after reading this article?
3. Make a list of those training programs needed and the frequency with which they should be held to correct the implied deficiencies in Fire Fighter McNamara's department.
4. Review a local fire department's training program and make an analysis of its training needs.

References

Firehouse. 2004 national run survey. *Firehouse,* June & August 2005, pp. 54 & 42, respectively, 2005.

Garvey, Gerald. *Public Administration: The Profession and the Practice.* New York: St. Martin's Press, 1996.

Goldfeder, Billy. Local "squeaking": The responsibility of firefighter training. *Fire Engineering Online.* http://fe.pennnet.com (accessed January 21, 2004). 2004.

NFPA. Evaluation and planning of public fire protection: Planning fire station locations. *NFPA Fire Protection Handbook,* 18th ed. Quincy, MA: NFPA, Chapter 10–20, 1997.

NFPA. *A Needs Assessment of the U.S. Fire Service. A Cooperative Study.* Authorized by U.S. Public Law 106–398: FEMA & NFPA. http://www.nfpa.org/assets/files/PDF/needsassessment.pdf (accessed July 24, 2005), 2002.

NFPA. *NFPA 1410, Standard on Training for Initial Emergency Scene Operations.* Quincy, MA: NFPA, 2000a.

NFPA. *NFPA 1500, Standard on Fire Department Occupational Safety and Health Program.* Quincy, MA: NFPA, 2000b.

USFA. *The Aftermath of Firefighter Fatality Incidents, Preparing for the Worst.* http://www.usfa.fema.gov/applications/publications/ (accessed July 24, 2005), 1997.

8

Health and Safety

Knowledge Objectives

- Comprehend the effect and importance of safety on fire and emergency services and its members.
- Realize the impact and influence that work-related injuries and deaths have on an organization and its members.
- Recognize the influence that the National Fire Protection Association (NFPA), U.S. Occupational Safety and Health Administration (OSHA), and other safety organizations have on the fire and emergency services.
- Examine the laws and regulations of state and local governments and their effect on fire and emergency services.
- Examine progressive safety trends in fire and emergency services.

Overview

The most valuable resource in any organization is people, and their well-being is vital to the organization's success. In a profession that is so dependent on physical skills, it is especially important to assure the health and safety of its members.

The issue of safety has been controversial at times. When the first safety standard for the fire service was issued by the National Fire Protection Association (NFPA) in 1987, many fire and emergency services leaders, along with city and county managers, prophesized that it would lead to departments going out of business or that it would take huge sums of money to comply with the safety standard. Neither of these predictions has come true. In many cases, fire and emergency services departments have used recommendations to acquire additional funding for important upgrades such as more modern protective clothing or increases in existing inadequate staffing.

Nevertheless, there seems to be selective compliance by some fire and emergency services agencies. Recently the NFPA reported, "The rate of fire fighter deaths at structure fires in the late 1990s was roughly the same as the rate in the late 1970s" (Fahy, 2002, p. 1). With the recent trend in fire-fighter deaths, it may be time for all fire and emergency services agencies to review their compliance with NFPA 1500, *Standard on Fire Department Occupational Safety and Health Program.*

NFPA Safety and Health Standard

NFPA 1500, *Standard on Fire Department Occupational Safety and Health Program,* on which most of this chapter is based, is relatively new. Many fire and emergency services organizations are in the process of adopting the requirements contained in this standard, but very few comply with it in its entirety. This is not an unusual situation for a newly developed trend in the fire service or in any organization.

NFPA 1500 was created to make the fire service profession less dangerous by reducing the risk of accidents, injuries, and fatalities in the course of conducting emergency operations and routine duties in the station. NFPA 1500 is also effective at helping fire fighters become healthier and more physically able to undertake their duties. This all has a direct benefit for the individual, the organization, and the public they serve.

Administrating the safety and health of the members of the organization should be one of the highest priorities for several reasons. The first is a moral obligation to provide a safe working environment for the members of this very hazardous profession (**FIGURE 8-1**).

Second, it is expensive to take care of fire fighters who are injured on the job. In the case of burns, the costs can be exorbitant and complete recovery can take years. Many fire and emergency services personnel will be better motivated if they know that their administration is looking out for their welfare. This can be a real morale booster.

In addition, many of the requirements of NFPA 1500 assure that the fire fighters will be competent professionals. This standard outlines minimum training and education along with physical fitness that provide the skills and knowledge needed to do an effective job as a fire fighter.

FIGURE 8-1 A Hazardous Profession

For example, over the past few years there has been a realization that although firefighting is very dangerous, many fire-fighter deaths and injuries are preventable. Many national leaders and organizations have attempted to influence the fire service to follow safety standards. Nevertheless, live fire training exercises with recruit fire fighters have been conducted where the instructors deliberately create unsafe, extremely high temperature evolutions. Clearly, the culture remains unchanged in some organizations.

This points to a fire and emergency services *safety paradox*. We make every effort to recruit members who are courageous and are risk takers. When faced with a dangerous situation, this type of person is looking for ways to rescue the victims and save their property from destruction, even at their own peril. We need these brave members. The challenge, through discipline and training, is to make these members safety-conscious and to make them practice safe behavior at emergency incidents, so they will take care of their own well-being and take risks only when necessary.

The number one cause of fire-fighter deaths, heart attack, is well documented. But many departments still do not require a medical exam to detect heart disease. If these individuals could be identified and either placed in a corrective medical and physical fitness program or removed from firefighting duties, their lives could be saved. But the culture of risking one's own life to save another's is so strong that even individuals for whom it is medically unsafe to perform strenuous physical work are sometimes even encouraged to participate. This issue is still a work in progress.

NFPA 1500

Even today, some of the items in NFPA 1500 are ill defined and there is no universal consensus on the exact meaning of the requirements. The following is a list of the major topics in NFPA 1500, along with a brief explanation of each:

- *Administrative:* The standard contains numerous items that require documenting the agency's organization and policies.
- *Safety committee:* This is a major part of implementing a comprehensive safety program. This committee provides the avenue to plan and implement changes needed to comply with the standard. It also provides a real-time discussion of current, potentially unsafe equipment; employee behavior; and operations. Members of this committee

should be selected from management, labor, and mid-level officers. This helps assure that discussions will focus on considering and assessing everyone's point of view. It is also useful to invite a representative from the municipality's risk management office. Besides providing the opinion of a safety expert, he or she will have extensive knowledge of the fiscal impact, including insurance costs and workman's compensation issues.

- *Standard operating procedures (SOPs):* These are normally developed to provide a safe and efficient action to manage a specific emergency operation such as a structural fire. The safety committee generally provides input to prevent deaths and injuries.
- *Training and education:* These issues are covered in Chapter 7.
- *Accident prevention:* Next to operating on the fireground, vehicle crashes are the most common cause of duty-related fire-fighter deaths. We will discuss this specific safety problem later in this chapter.
- *Equipment and vehicles:* Most of these issues have been addressed by a majority of fire and emergency services organizations. The federal government's Fire Grant program is helping many needy departments purchase newer, safer apparatus and equipment. Unfortunately, rural and other small departments are struggling to find enough money to buy safety equipment.
- *Health and wellness:* This issue is still in the process of consensus building among fire experts. Consistently noted as the number one cause of fire-fighter deaths, we have made little progress in addressing this issue, and the requirements in NFPA 1500 are sometimes unclear. We will discuss this safety issue in greater detail later in this chapter.
- *Incident management system:* Progress in this area is ongoing. Even after the terrorist incidents on September 11, 2001, the New York City police and fire departments have not come to an agreement about who should be in charge at the next major event. ". . . they [NYC police and fire commissioners] flatly rejected the concept of a unified, integrated incident management system for large-scale incidents in New York City" (Manning, 2002, p. 4).

Fire Apparatus Crash Prevention

Each year, motor vehicle–related incidents account for a significant share of U.S. fire-fighter deaths. According to NFPA statistics, from 1987 through 1996, 272 U.S. fire fighters died in such incidents. Of these 272 deaths, 127 were the result of crashes. Most of these crashes occurred while fire fighters were responding to or returning from alarms. Forty-nine of the crash deaths involved fire department apparatus. (Sixty involved fire fighters' personal vehicles and 15 involved aircraft.) (NFPA, 2005, p. 20)

In a *Detroit News* special report about ambulance safety the newspaper reported ". . . dangerous driving by rescue workers endangers the lives of the sick and injured and those sent to help them. When medics settle into the back of an ambulance with their patients, they might reach for the oxygen, medicine or paperwork, but they aren't likely to reach for their seat belts" (Zagaroli, 2003).

"Firefighters are more likely to die traveling to or from a fire than fighting one, and motor vehicles pose a greater hazard than flames, according to new data from the National Fire Protection Association (NFPA). All told, 105 firefighters died while on duty in 2003, up from 97 in 2002 . . ." (NFPA, 2004).

These vehicle crashes are tragic by themselves, but the total impact on the fire service also includes fire-fighter injuries and deaths, liability issues, and out of service time for fire apparatus. A vehicle accident can cause both a financial and an operational disaster for the department.

The other side to the vehicle accident deaths, injuries, and property damage problem is the civilian vehicles that are struck by fire and emergency services vehicles. In many cases, emergency vehicle drivers are risking the lives of unaware civilian drivers and their passengers when they drive too fast for conditions, fail to fully stop at controlled intersections, and place too much dependence on their emergency lights and siren to clear traffic.

For example, Vincent Brannigan, in a July 1997 article for *Fire Chief* magazine, reported that "In March, a . . . City fire truck on its way to a fire hit a small car containing six young people, killing three and seriously injuring the rest. Such an accident could easily result in damages against the city exceeding $10 million" (Brannigan, 1997).

Paradoxically, fire fighters will risk their own lives to rescue a victim from a fire, but while responding to that same fire will unknowingly place themselves, fellow fire fighters, and civilians at risk while aggressively responding to the emergency call. And what makes many of these crashes even more tragic is that a good number of the calls turn out to be nonemergencies.

NFPA 1451, *Standard for Fire Service Vehicle Operations Training Program,* Chapter 3, Training and Education, contains the training requirements for defensive driving techniques. NFPA 1002, *Standard for Fire Apparatus Driver/Operator Professional Qualifications* and NFPA 1500, *Fire Department Occupational Safety and Health Program,* Section 4-2, Drivers/Operators of Fire Department Vehicles, contain guidelines for the basic training new drivers should receive before being permitted to operate fire department vehicles. These are all key elements in reducing accidents. It is always easier to learn a new skill correctly than to change existing habits. Nevertheless, existing drivers should be encouraged to complete a driver-training course designed to comply with these safety standards.

In-service training for existing drivers is required to be given at least twice a year. This provides valuable hands-on experience and training to drivers who are rarely behind the wheel. For all drivers, this is an opportunity to reinforce defensive driving techniques and review emergency response SOPs.

Vehicle Operation Laws and Liability

States have provisions in their vehicle driving regulations that allow emergency vehicles to exceed posted speed limits and proceed through controlled intersections without coming to a complete stop. These emergency driving regulations contain some language that assumes the emergency vehicle driver will use all due caution in being granted this privilege. If the emergency vehicle hits another vehicle while exceeding the speed limit or proceeding through a stop signal or sign, the driver was likely driving with some carelessness.

> . . . the [emergency] vehicle must normally be driven with reasonable care under the circumstances." This is the really tricky part. Courts recognize that reasonable care in driving a fire truck is not the same thing as reasonable care in driving a car. However, the general responsibility to drive in such a way that you don't cause accidents is unchanged. . . . Notwithstanding these definitional differences, it is improbable that the driver of an emergency vehicle approaching a controlled intersection on a red light might not foresee that there would be cross traffic moving through the intersection. It is the foresight of the specific dan-

> ger which makes this accident avoidable with the exercise of ordinary care. (*Schwarzbek v. City of Wauseon [Ohio]*, August 23, 1996, as cited in Brannigan, 1997, pp. 28-29)

Provisions to control the most common causes of fire service serious accidents are outlined in NFPA 1500. For example: Drivers shall come to a "complete stop and shall not proceed until it is confirmed that it is safe to do so" is listed for situations such as any stop signal or sign, a stopped school bus, and unguarded railroad crossings (6-2.8). Written policies (SOPs) governing vehicle speeds and limitations during inclement weather are needed to reduce accidents and any potential liabilities (**FIGURE 8-2**). A common limitation is that the speed be no more than 10 mph above the posted speed limit. It is important that the policy be very specific rather than simply stating "as appropriate" or "safely."

Again, Brannigan quotes from a Pennsylvania Supreme Court Justice as follows:

> But it is to be noted in this connection [fire department accident] that while the law—in view of the vital and urgent missions of fire department vehicles—wisely allows them certain privileges over other vehicles, it does not assign to them absolute dominion of the road. It is unfortunate that the very attribute which sounds the highest praise for fire companies, namely, speed, should have been the very demon which brought about their undoing. . . . But speed, which is uncontrolled, is capable of wreaking as much havoc and causing as much sorrow as fire itself. (Brannigan, 1997, pp. 28-29)

NFPA 1500 also contains other items that will help prevent accidents and injuries and should be included in the SOPs. For example, when operating at a vehicle accident on a highway, members should wear a garment with fluorescent reflective material for visibility, use fire apparatus as a *shield,* and place appropriate warning devices including highway caution signs to warn oncoming traffic.

The inspection, maintenance, and repair of vehicles has been identified as a method for preventing several serious fire service vehicle accidents in the past. Requirements for creating a program to eliminate these causes are outlined in NFPA 1500, Chapter 8, Vehicle and Apparatus Care.

FIGURE 8-2 Governing Vehicle Limitations to Reduce Accidents and Potential Liabilities

Safety Priorities

In general, the main causes of fire-fighter deaths are not a secret, nor are they new. The fire-fighter fatality reports for the last 20 years have clearly shown that heart disease and vehicle accidents account for approximately 64%–69% of the fire-fighter fatalities in any one year. Although all the other items in the safety standards are important and do save lives and prevent injuries, these two specific problems have the potential to save the greatest number of fire-fighter lives.

> Let me start by saying that there seem to be a lot of people dropping over because of heart attacks. This does not surprise me, as I have lived the lifestyle of the brave and cholesterol ridden for many years. The causes of this malady are well known. You eat too much, you smoke too much and you don't exercise at all. And then suddenly you are racing to the scene of a blaze, raising ladders, dragging hose and rescuing victims. (Carter, 2001)

Many vehicle accidents are preventable. Safety professionals dislike calling them accidents and prefer to call them crashes. They are generally not accidents, because most could be prevented. Most of the serious vehicle crashes are the results of excessive speed, failure to stop at traffic signals and stop signs, and general lack of caution at intersections.

Finally, accident investigation and analysis has been incorporated into the standard. Without feedback, this safety prevention program would not be able to adjust to new hazardous situations, information, and equipment.

The following is a summary of a recent study on fire-fighter deaths:

Heart Attack Leading Cause of Death for Fire Fighters

WASHINGTON, D.C.-The United States Fire Administration (USFA) released today a comprehensive study which examines the causes of deaths for "on-duty" fire fighters. The USFA *Fire Fighter Fatality Retrospective Study: 1990–2000* is an in-depth analysis as to the causes for more than 1,000 on-duty deaths which have occurred in the United States during the last decade of the 20th century. The goal of the study is to identify trends in fire fighter mortality, and use the information to help reduce fire fighter deaths by 25% in 5 (five) years.

The key findings of the study include:

- The leading cause of death for fire fighters is heart attack (44 percent). Death from trauma, including internal and head injuries, is the second leading cause of death (27 percent). Asphyxia and burns account for 20 percent of fire fighter fatalities.
- Each year in the United States, approximately 100 fire fighters are killed while on duty and tens of thousands are injured. Although the number of fire fighter fatalities has steadily decreased over the past 20 years, **the incidence of fire fighter fatalities per 100,000 incidents has actually risen over the last 5 years** [emphasis added by author], with 1999 having the highest rate of fire fighter fatalities per 100,000 incidents since 1978.
- Fire fighters under the age of 35 are more likely to be killed by traumatic injuries than they are to die from medical causes (e.g., heart attack, stroke). After age 35, the proportion of deaths due to traumatic injuries decreases, and the proportion of deaths due to medical causes rises steadily.
- Since 1984, motor vehicle collisions (MVCs) have accounted for between 20 and 25 percent of all fire fighter fatalities, annually. One quarter of the fire fighters who died in MVCs were killed in private/personally owned vehicles (POVs). Following POVs, the apparatus most often involved in

fatal collisions were water tankers, engines/pumpers, and airplanes. More fire fighters are killed in tanker collisions than in engines and ladders combined.

- About 27 percent of fatalities killed in MVCs were ejected from the vehicle at the time of the collision. Only 21 percent of fire fighters were reportedly wearing their seatbelts prior to the collision.
- Approximately 60 percent of all fire fighter fatalities were individuals over the age of 40, and one-third were over the age of 50. Nationwide, fire fighters over the age of 40 make up 46 percent of the fire service, with those over 50 accounting for only 16 percent of fire fighters. About 40 percent of volunteer fire fighters are over the age of 50, compared to 25 percent of career fire fighters.
- The majority of fire fighter fatalities (57 percent) were members of local or municipal volunteer fire agencies (including combination departments, which are comprised of both career and volunteer personnel). Full-time career fire fighters account for 33 percent of fire fighter fatalities. Numerically more volunteer fire fighters are killed than career personnel, yet career personnel lose their lives at a rate disproportionate to their representation in the fire service.
- In many fire departments, EMS calls account for between 50 and 80 percent of their emergency call volume. These EMS incidents result in only 3 percent of fire fighter fatalities. Trauma (internal/head) accounts for the deaths of 50 percent of fire fighters who were involved in EMS operations at the time of their fatal injury. Another 38 percent involved in EMS operations died from heart attack.

For the past 25 years, the United States Fire Administration (USFA) has tracked the number of fire fighter fatalities and conducted an annual analysis. Through the collection of this information on the causes of fire fighter deaths, the USFA is able to focus on specific problems and direct national efforts to finding solutions for the reduction of fire fighter fatalities in the future. The information in this study is also used to measure the effectiveness of current programs directed toward fire fighter health and safety. One of the USFA's main program goals is a 25 percent reduction in fire fighter fatalities in 5 years and a 50 percent reduction within 10 years. (USFA, 2002)

For stress/heart-related fire-fighter deaths (44% of all fire-fighter deaths), a two-pronged approach works best. First, medical exams as outlined in NFPA 1500 and 1582 should be scheduled for all members. Any indication of heart disease should be treated without delay. In most cases, the member should be completely evaluated before being allowed to participate in emergency operations duties.

Second, an ongoing physical fitness program has many beneficial outcomes. This type of training reduces the chance of a heart attack, lowers cholesterol, reduces body fat, and strengthens the heart. Also, those persons with stronger hearts as a result of aerobic exercise are more likely to survive a heart attack.

Physical Fitness

A national consensus (for example, an NFPA standard) on how to measure physical fitness competency and minimum acceptable level has not been formulated at this time. Therefore, each fire and emergency services organization needs to find a competent physical fitness professional to recommend and monitor this type of program. Several legal and ethical issues also need to be resolved before implementing a physical fitness program.

In general, if tests are based on actual job performance tasks and the same tests are used for both applicants and incumbents, these tests will be able to be defended in court. Any adjustments for age or sex will fail the test of treating all fire fighters equally. The only possible exception to this rule of thumb is *percent fat calculations*, where there is justification for females to have higher percentage levels.

Actually, there are some minor costs. Fire fighters who exercise regularly are able to use their air supplies more efficiently, allowing them to spend more time on-air. The improvement will vary depending on how out-of-shape aerobically the individual is when they begin an exercise program. If the body is trained to average levels of aerobic conditioning, it has the ability to use oxygen more efficiently.

A Common Sense Approach to Fire-Fighter Physical Fitness

Physical fitness has long been recognized as an essential trait for fire fighters. The profession requires a high level of strength and endurance, and many tasks cannot be done properly without it. An adequate level of physical fitness is a "safety" goal both for the public served and for the individual. The following are some important points on physical fitness in the fire-fighting profession:

- A fire fighter in good aerobic condition can operate for a longer period of time with a SCBA than a fire fighter who is in poor aerobic condition. Fire fighters in good physical condition will have a more efficient use of their air from the SCBA, which will allow them to work for longer periods of time in a hostile atmosphere. This becomes critical when a fire fighter with a poor aerobic capacity has to leave after their low air alarm activates, requiring the withdrawal of the team.
- Good levels of physical fitness provide two benefits: first, the public will receive outstanding performance from their fire fighters; second, the individual fire fighter will have a better chance of surviving or preventing a heart attack or stroke.
- Heart attacks and strokes are the number one killer of fire fighters. In an interview, Battalion Chief Pfiefer, FDNY, stated, "There were messages, urgent messages of fire fighters having chest pains as they went up . . ." (*Firehouse*, 2002, pp. 42–44) as he spoke about his experiences as the first due battalion chief at the World Trade Center on September 11. Another FDNY officer reported "As we got higher into the building, we heard numerous 'Urgents' and 'Maydays' from fire fighters with chest pains, in need of oxygen or worse." (*Fire Engineering*, 2002).
- Being overweight: "It's a problem that tarnishes our image and jeopardizes our rela-

tionship with the public. It reduces our effectiveness in handling emergencies, and it threatens our health and safety" (Page, 2002).
- The public has a right to expect that the fire fighters who are responding to their emergency are trained, organized, staffed, equipped, and in top physical shape. Anything less is giving your profession, your fellow fire fighters, and the public less than your best effort.
- A fire in a building does not know if the fire fighters attempting to extinguish it are male or female or their age. Physical fitness should be directly related to the job, not gender or age.
- Fire fighters work in teams. Each one should be expected to be able to help their partner if they go down inside a building on fire. No exceptions!

Without proper physical training, the human body loses strength and endurance predictably as we age. Nevertheless, with proper physical training a person can function at high levels of strength and endurance into his or her 60s.

The following items can help guide the evaluation of a department's physical fitness program. Mandatory physical fitness programs can make huge improvements.
- When deriving physical fitness minimum levels, the needs of the public (our customers) and fire-fighter safety must come first.
- Validating physical fitness by testing incumbents is sometimes inaccurate. There will be incumbents who are performing poorly because they do not participate in a physical fitness program. Most have the potential of doing a lot better with training. Also, because there was no physical fitness standard when they were hired or accepted as a member, some incumbents may be unable to perform all the essential job functions.
- A comprehensive wellness program that includes medical evaluations must be used in combination with an exercise program, and the costs should be borne by the department.
- When first implemented, a physical fitness standard for incumbents should recognize that some fire fighters have not exercised for many years, and so for fairness, should be allowed a phase-in period to meet standards. Most fire fighters who find themselves identified as below standards for a physical fitness evaluation will be able to meet standards with the appropriate training.

- A fire fighter who fails to meet the minimum levels of physical fitness (different levels for incumbents during the phase-in period) should be prevented from participating in physically demanding work. In the *Fire Department Occupational Health and Safety Handbook,* Stephen Foley writes "Members must not be allowed to begin or resume suppression duties until they pass the physical performance requirements" (Foley, 1998, p. 87).
- A voluntary or nonpunitive physical fitness program guarantees that there will be members in poor physical fitness. For example, if a traffic intersection in your town has stop signs posted to prevent accidents, and the local ordinance is voluntary or nonpunitive, there will be drivers who disregard the stop sign. This guarantees some serious accidents. Safety and health requirements must be mandatory in order to work.
- If your department purchases a new 1,000 gallons per minute (GPM) pumper, and 15 years later you test the pumper for flow and find it is now only 750 GPM as a result of use and a lack of proper maintenance, would you take action to fix this vital piece of firefighting equipment?
- If your department hires or accepts a new member at a physical fitness of 100%, and 15 years later that person operates at 75%, what would be the fair thing to do from the public's perspective?

Federal Safety Regulations

The U.S. Constitution has language preventing the federal government from having too much power over the states, so many federal regulations are not enforceable at the state or local level. In the case of U.S. Occupational Safety and Health Administration (OSHA) regulations, when first adopted they were not enforceable at the state or local level. In those states that decided to have a state plan, however, the individual state adopted the federal OSHA regulations and they became enforceable for state and local government employers and employees.

The following states have OSHA-approved plans: Alaska, Arizona, California, Connecticut, Hawaii, Indiana, Iowa, Kentucky, Maryland, Michigan, Minnesota, Nevada, New Mexico, New York, North Carolina, Oregon, South Carolina, Tennessee, Utah, Vermont, Virginia, Washington, and Wyoming. In addition, Florida, Illinois, and Oklahoma have typically adopted OSHA regulations for use by fire departments.

Nonparticipating states may or may not adopt the federal safety requirements for state and local government employees. In addition, many departments use the federal and NFPA safety standards with the understanding that they are providing a safe working environment for their members and this policy reduces any potential liability.

Joint Labor-Management Wellness/Fitness Initiative

This is a national-level initiative of the International Association of Fire Fighters (IAFF) and the International Association of Fire Chiefs (IAFC) to establish a program for medical and fitness evaluations, rehabilitation, behavioral health, and data collection. One of the basic premises of this effort is that it is not designed to be *punitive.* This can be in conflict with the belief that if a member is not fit for duty based on either medical or physical fitness criteria, he or she should be prevented from participating in emergency operations. Is this a *punitive* action? The assumption is that in a nonpunitive situation, they would be placed on light duty, injuries on the job (IOJ) leave, or some other placement where there would be no reduction in wages or loss of their jobs.

The next phase for a full salaried, "unfit" fire fighter is placing them in rehabilitation. How many salaried fire fighters could a department afford to place in non-combat positions before it would become a hardship? How many volunteers could be taken off combat status before it would adversely affect responses? This is a complex and highly individualized problem that could have a substantial impact on the budget and operation of the department. If an appropriate, reasonable implementation plan is used, however, the number of fire fighters needing rehabilitation will probably be small.

When a new program is initiated, it is reasonable to allow fire fighters who have medical and physical fitness problems an equitable time frame in which to participate in a rehabilitation program. Also, NFPA 1500 allows the phased implementation of safety requirements. For example, if a fitness test was chosen that has a pass-fail time of 7 minutes, the first year could start out with a time of 13 minutes for incumbents. Then each year, reduce the time by 1 minute until all members meet the 7-minute standard.

The Rest of the Big Picture

There are many other safety items identified in NFPA 1500 that should be addressed. In addition, the safety committee should review the annual *U.S. Fire Fighter Fatalities and Injuries* reports issued by the NFPA. Also, the *USFA Fire Fighter Fatality Retrospective Study: 1990–2000* identifies the major causes from a multi-year perspective. These reports contain many clues to areas where deaths and injuries can be prevented.

Safety Facts and Studies

Many diverse safety studies have been done on fire and emergency services members. Here are a few examples:

- *National Fire Fighter Burn Study (2002):* This study was done in association with the Washington Hospital Burn Center, Washington, D.C., and the Maryland Fire and Rescue Institute, University of Maryland, College Park, Maryland. In-patient fire fighters thermally injured in the line of duty were interviewed by questionnaire while in the hospital by burn center staff, and at 6 and 13 months after discharge by burn study staff. Outpatients completed the questionnaire prior to leaving the hospital. Participation in the study was voluntary and confidential. No information shared in the interview would be disclosed to the fire fighter's department. The data for this study were collected only from fire fighters treated at 20 burn centers throughout the United States. The following are some of the results:

- 22% were required to wear glasses but were not wearing them.
- 16% had not achieved Fire Fighter I competency.
- 22% received burns at a training fire.
- 6% were alone when burned.
- 29% were standing in the fire area.
- 6% had consumed alcohol within 12 hours.
- 64% were advancing a hose line.
- 59% described the heat buildup as "instant."
- 24% were in a flashover.
- 20% were not wearing a hood.
- In 18% of the hand burns the fire fighter did not have gloves on.
- 88% reported that they were "reluctant to return" to duty.
- 45% felt the injury was preventable.
- Among the findings of a University of Minnesota's Carlson School of Management/City of Minneapolis study (Liao et al., 2001) are the following:
  - Fire fighters who tended to ignore safety rules and regulations had accidents more frequently and suffered more severe injuries than fire fighters who conscientiously followed safety rules.
  - Female fire fighters reported 33% more injuries than male fire fighters reported.
- A study published in the *Journal of the American Dietetic Association* says that of 96 full-time fire fighters surveyed in a U.S. city (un-named), 84% were found to be overweight (13% were obese) (Kay, 2001).

CASE STUDY: Injuries on the Job

Approximately 22% of all fireground injuries are categorized as "wounds, cuts, bleeding or bruises." Because there are a large number of injuries in this category, a program could be initiated to prevent these injuries. Training and education would be a part of the overall plan, but the department's officers would also want to be sure that all members are wearing all their safety equipment and are following SOPs.

One important item for the prevention of cuts, wounds, burns, and other injuries of the hand is the use of issued fire-fighter safety gloves. If a safety officer receives an injury report of a cut to the palm of the hand from a fire fighter, he or she would want to examine the glove of the hand that was cut. It should have an identical cut and blood around the cut in the glove. Without the matching cut in the glove, it could be inferred that the fire fighter was not wearing gloves when the injury occurred, and corrective action would be taken. There may be times when there is a suspicion, but it cannot be proven beyond a shadow of a doubt. Even in these cases, if the members know the administration is checking, they will be more likely to wear all their personal protective clothing.

Discussion Questions

1. Is it fair to check up on the fire fighters as if they were 2-year-olds?
2. Should the union file an unfair labor practice complaint against the chief?
3. Would it be appropriate to have the fire fighter use their own health insurance rather than the jurisdiction's worker's compensation if the fire fighter was careless or neglected to follow safety SOPs or wear safety equipment?
4. Is this an example of managing by walking around? If yes, are there other actions that could be taken using MBWA that would prevent injuries?
5. How would the administration be assured that all fire fighters know what safety rules, including the use of safety equipment, they should use during emergency operations?

References

Brannigan, Vincent. The real hazard is driving the fire truck. *Fire Chief,* July 1997, pp. 28–29.

Carter, Harry R. Why are we dying? *Firehouse,* August 2001, p. 21.

Dwyer, Jim, et al. Fatal confusion: A troubled emergency response; 9/11 exposed deadly flaws in rescue plan. *New York Times,* July 7, 2002. Metropolitan Desk, p. 1.

Fahy, Rita F. *U.S. Fire Service Fatalities in Structure Fires, 1977–2000.* Quincy, MA: National Fire Protection Association, 2002.

Fire Engineering. Anonymous letter. *Fire Engineering,* September 2002, p. 97.

Firehouse. WTC—This is their story. *Firehouse,* April 2002, pp. 42–44.

Foley, Stephen N. (Ed.) *Fire Department Occupational Health and Safety Standard Handbook.* Quincy, MA: NFPA, 1998.

Kay, Brigett F., et al. Assessment of firefighters' cardiovascular disease-related knowledge and behaviors. *Journal of the American Dietetic Association, 101*(7), p. 807, 2001.

Liao, Hui, Richard D. Arvey, & Richard J. Butler. Correlates of work injury frequency and duration among fire fighters. *Journal of Occupational Health Psychology, 6*(3): pp. 229–242, 2001.

Manning, Bill. Blood on their hands—Editor's opinion. *Fire Engineering,* November 2002, p. 4.

National *Fire Fighters Burn Study.* http://www.fire fighterburnstudy.com (accessed June 10, 2003). 2002.

NFPA. Vehicle crashes cause more firefighter deaths than fires. *NFPA News Release,* June 9, 2004. Quincy, MA: NFPA.

NFPA. *Selected Special Analyses of Firefighter Fatalities.* http://www.nfpa.org (accessed July 24, 2005), 2005.

Page, James. A taboo topic: Fire service professionals ignore a big problem. *Fire-Rescue Magazine,* October 2002, p. 10.

USFA. *Heart Attack Leading Cause of Death for Fire Fighters.* http://www.usfa.fema.gov/about/media/2002 releases/02-193.shtm (accessed September 24, 2005), 2002.

Zagaroli, Lisa. Ambulance trips put lives at risk. *Detroit News,* January 26, 2003, Front page.

9

Government Regulation, Laws, and the Courts

Knowledge Objectives

- Comprehend the creation, effect, and importance of government regulations and laws.
- Identify and understand the moral and ethical justifications for government oversight of the public fire and emergency services.
- Understand the impact and influence that the legal system and the courts have on fire and emergency services.
- Recognize the influence that the National Fire Protection Association (NFPA) has on the fire and emergency services.
- Examine and comprehend building codes and regulations at the state and local level and their effect on fire and emergency services.
- Examine progressive laws and regulation trends in fire and emergency services.

Government Regulations

What kind of impact do regulations, laws, and standards have on fire, EMS, and emergency services organizations? In a *Wall Street Journal* article on February 7, 2001, Robert Johnson predicted that a National Fire Protection Association recommendation (NFPA 1710) "could prompt fire departments to hire 30,000 firemen nationwide, an 11% increase" (Johnson, 2001). If true, this would have a substantial fiscal impact on our nation's fire and emergency services and taxpayers. However, without getting into specifics, the impact estimated by Johnson was not substantiated by any scientific study nor was any fiscal analysis included in the article. Where and how are these regulations and standards created and by whom?

At the local level, city/county administrators and elected officials along with some fire and emergency services officials protest unfunded mandates, which occur when a government regulation requires the expenditure of tax dollars at the state or local level but is unaccompanied by money for the implementation. For example, when the U.S. Department of Transportation issues new safety design features for roads, it is not uncommon for the federal government to provide matching funds to encourage states to comply. This is a partially funded government mandate. Most are never fully funded.

This is a complex subject. A good place to start discussing this topic is with an explanation of the justification for these documents that become the *standard of care* for the fire and emergency services professions.

Justification for Government Regulation

When is it justified for a government to regulate private citizens or businesses in a free country? Individuals, businesses, and organizations that become the target of government regulation commonly resist and protest that there is no need for government intervention in their affairs. Most fire and emergency services providers are either a part of local or state governments or independent volunteer fire companies, and their elected officials and appointed administrators will argue that the federal or state government has no business telling them what to do. They protest that standards of service are best determined at the local level of government.

In general, there are three major reasons why government regulation can be justified. First, it is the rightful role of government to protect the consumer in cases where it can be demonstrated that the consumer lacks the knowledge, background, or education to make a conscious judgment of the safety, competency, or adequacy of the service or product. Government regulation is usually seen as justified for the public when the everyday consumer cannot evaluate the product or service with information or knowledge normally available to them.

As an example, governments typically regulate the contents and labeling of food products. It would be impossible for the consumer to know the health risks or contents of, for example, a can of vegetables without first sending it to a lab to be tested. The consumer would have no readily available knowledge about how the vegetables were planted, grown, harvested, processed, and canned. But because of government health regulations, a consumer can be assured that the contents are accurately described on the label and are acceptable for consumption. Without some assurance of the safety of this food product, few consumers would continue to buy the product. Therefore, this regulation benefits both the consumer and the producer.

Also, the state or federal government regulates most professional services such as doctors, lawyers, and engineers. It would be impossible for most consumers of these professional services to accurately assess the competency of these professionals. The customer would have to know the college they attended and if the academic standards were sufficient to meet a minimum standard of education. Currently, these professions require a competency exam to prove test-takers' mastery before they are granted a license to practice in their state. When consulting a doctor or seeking advice from a lawyer, the client must exercise a certain level of trust. Minimum license requirements assure the public that the professional they are dealing with has basic knowledge and competency regarding the profession.

Second, a market failure may prevent a safety product or service from becoming available to the general public. Typically, market failures are more common in those products or services that can prevent injury, sickness, or death. For example, seat belts are required by government regulation in all vehicles sold in this country. At the time the seat belt regulation went into effect they were offered as an option, but very few car buyers chose them because they were uncomfortable and the likelihood of being involved in a car crash seemed low. Thanks to the perseverance of safety groups such as the American Automobile Association and insurance associations, the federal government issued a regulation requiring seat belts in all newly manufactured cars.

Once seat belts were required in cars, it was assumed that drivers and the occupants would use them. But there was resistance, so states, encouraged by insurance companies and safety organizations, passed laws requiring the wearing of seat belts. In most cases, the regulation requiring safety equipment must be paired with rules that require its use and provide some type of disciplinary action for noncompliance. This same type of justification is behind the mandatory installation and maintenance of smoke alarms in homes.

Finally, when a product or service becomes a monopoly, the consumer loses the ability to choose a product or its price. Some typical consumer monopoly products are telephone service, electricity, water, and sewer. Private companies or the government can provide these, but they all work most efficiently when granted a monopoly regulated by government.

For example, imagine that telephone service has four separate providers that would compete for the public's business. Every telephone pole in the neighborhood would have four separate wires. Because of the heavy costs for the infrastructure, the government grants a monopoly and controls both quality and price.

For another illustration, let's imagine again that the interstate roads were actually part of the competitive open market. When leaving to go on vacation or a business trip, the traveler has the option of choosing one of several road providers, similar to choosing an airline. Each road system may have its own quality of pavement, speed limits, and tolls. Again, because of the extreme costs to build roads, they are almost all provided as a public good by the government. Drivers are required to pay a user fee (tax) that is included with the cost of fuel. There is no choosing because this is a monopoly funded and controlled by the government.

Is Government Justified in Regulating Fire and Emergency Services Agencies?

The following questions can be used to determine if there is justification for regulating fire and emergency services agencies:

- Is the customer knowledgeable about the product or service? When the fire engine shows up at a house fire, can the typical customer determine the competency of the fire fighters and whether there are enough of them, the adequacy of the fire equipment, and the organization and supervision of the crews? To become a professional, a competent fire fighter requires extensive time in a specialized training program for the mastery of the skills and knowledge necessary to become proficient.
- Is there a market failure for a public good? If the consumer were to purchase public fire protection, would they pay or contribute before actually needing this critical emergency service? A small number of organizations do fund their operations by voluntary donations, but this situation is found only in small communities. Studies have shown that most people do not believe they will experience a hostile fire, especially in their homes. This is a classic example of a public good that in most cases cannot be withheld from anyone who requests emergency assistance. Even those who do not pay or contribute will receive the service.
- Is there a monopoly? Clearly yes, because the public has only one selection when dialing 9-1-1. This is another situation that would be very expensive if there was more than one provider. Because response time is such a critical part of emergency service, a given system can provide reduced response time by having all stations be part of the same company rather than having two or more stations servicing neighborhoods. Imagine the absurdity of a billboard announcing, "We are the best! You light them, we fight them. Call us for your next fire."

Although regulations and standards for local fire and emergency services are relatively new in this country, there seems to be adequate justification for the protection of the public to expect and encourage government oversight.

Federal Regulation

At the federal level of government, laws are adopted through a process that involves the legislature passing laws and the president approving them. This is a generalization, however; there are several variants, such as when the president vetoes a bill.

Generally, Congress does not put specifics for implementation into the regulatory legislation. They usually defer to the government bureaucracy to formulate the "rules" to be followed to comply with the new or revised law. This is done for two practical reasons.

First, the legislature normally lacks the expertise of the professional bureaucracy. For example, the U.S. Congress passed and the president approved a public law that created the Occupational Safety and Health Administration (OSHA). The law contained guidelines and a process for adopting regulations. This law had a great impact on employees and employers, as can be gleaned from the following summary extracted from OSHA's Web page (OSHA, 2004):

> OSHA's mission is to ensure safe and healthful workplaces in America. Since the agency was created in 1971, workplace fatalities have been cut in half and occupational injury and illness rates have declined 40 percent. At the same time, U.S. employment has doubled from 56 million workers at 3.5 million worksites to 111 million workers at 7 million sites.

Second, lawmakers do not like to place themselves in the middle of controversial regulations that may end up "lose–lose" for them. Typically, they like to spend tax revenue and bring back federal projects to their districts. They can take credit for this spending in their districts, but there is no political advantage to adopting new regulations.

In the regulatory process, there are typically two groups that have a stake in the outcome. In the case of an OSHA safety regulation, there are the employees who are interested in their safety at any cost, and there are the employers who are interested in making a profit by keeping costs as low as possible. These are rarely black and white decisions, but instead have a lot of gray. Because of the complexity and the conflicts that arise with a proposed regulation, many take a lengthy period of time for review, public comment, and final adoption. For example, the Respiratory

Protection Rule, OSHA, 29 CFR 1910.134, took about 13 years from the first draft until the final rule.

A new process may help shorten this review and adoption process for the federal government. Executive Office of the President, Office of Management and Budget, Circular No. A-119 ". . . directs agencies to use voluntary consensus standards in lieu of government-unique standards . . . " (OMB, 1998). This was issued on February 10, 1998, and has the effect of directing federal agencies to use voluntary consensus standards, such as the ones promulgated by the NFPA. The voluntary consensus standards organization is described as:

- Being open process
- Supporting a balance of interest
- Maintaining due process
- Providing an appeals process
- Working toward consensus, that is substantial agreement, but not necessarily unanimous support

Because of this relatively recent proclamation, and the extended time it takes for federal rule making, many existing regulations will still be based on requirements created by the federal bureaucrats. But this new executive order is very clear that federal agencies must use voluntary consensus standards in lieu of government-unique standards in their procurement and regulatory activities, except where inconsistent with law or otherwise impractical.

Democratic Accountability versus Administrative Discretion

In the United States, some would like administrative civil servants to be directly accountable to the public. But we have a representative form of government, where officials are elected by majority vote of the public and become the representatives of the voters. This is actually a representative form of government, not a democracy. It is common for these representatives to create laws that exist mainly to ensure that appointed officials follow their guidance.

What seems to work best is a division where the policy decisions are the responsibility of the elected officials and the implementation plans are the responsibility of the technically competent administrator. For, example, the elected officials decide that paramedic service is something the municipality should provide for the community. For this example, the assumption will be made that the department has more than one station, and it would be up to the fire or EMS officials to propose how many units are needed

and their exact locations. In other words, they would maximize the service level given a specific budget and general direction by elected officials.

Controversial Policy Decisions

There are times when a policy decision has a lot of political baggage. This happens with controversial topics that have strong advocacy groups and many potential winners and losers. "Public administrators not only would have internalized the values of the citizenry in general but also would have committed themselves to the highest moral and practical standards of their profession" (Garvey, 1996, p. 128). Therefore, the public administrators would be best qualified to judge the fairness and appropriateness of a rule and its effect on the public.

When faced with political pressure from a special interest group, and the lack of real input from citizens, chiefs will have to rely on their ability to argue for a policy or decision change on their own. This may take every ounce of courage, honesty, perseverance, and knowledge that can be gathered. Remember, the primary advocate for the customers is the administration. There truly is no one else who can impartially represent the public's interest.

Chief administrative officials of a fire and emergency services organization should represent the public in any discussions of policy that affect service level and quality. They are the only persons who can focus on quality service rather than on other issues such as improving the benefits of a special interest group at the expense of service to the general public, although there will be times when improving the salary and benefits of fire fighters is the right thing to do.

It is in these times and situations that the chief will use the best negotiating and consensus building techniques possible to come to a fair and equitable solution. Try to keep the discussions in a public forum where the administration can have a fair chance to respond to any concerns of the special interest group. Face-to-face meetings can be very productive.

Administrative Rule Making

Administrative rule making allows the public administrator to function as a lawmaker. "The opportunity for administrators to function, in effect, as lawmakers often arises because the legislators themselves may lack the interest, the information, or the political courage to settle every last provision of a new public program" (Garvey, 1996, p. 138). For example, recently the Florida legislature adopted a state-wide minimum building code and fire prevention code. However, the legislature did not specify the exact model building code to adopt, but instead delegated that authority to the Florida Building Code Commission. This commission was made up of government representatives, subject matter experts, and stakeholders. It had to report its final determination to the legislature, but it would take some type of formal action for the legislature to overturn the commission's final decision.

The Florida state fire marshal was given the authority to adopt, modify, revise or amend, interpret, and maintain the Florida Fire Prevention Code. Although the state legislature specified the use of NFPA 1, *Uniform Fire Code*, and NFPA 101, *Life Safety Code*, as the two base codes, the administration and interpretation is delegated to the state fire marshal.

In the federal arena, OSHA is constantly developing and issuing new rules. "The rule maker's role—one of the most important in public administration—requires administrators to function as lawmakers, or at least in a quasi-legislative mode. That mode casts appointive officials in a role which strict democratic theory once reserved for elected representatives sitting in formal legislative bodies" (Garvey, 1996, p. 139). Usually these "administrative rules" are issued under the signature of the head administrative official such as the secretary of labor or the state fire marshal. However, generally a special team of experts within an agency and/or a task force of stakeholders will be assigned to see rule making through from start to finish and to make a recommendation to the administrative official.

Many meetings and numerous hours will be spent in consultation and negotiation with individuals and representatives of groups whose members will be affected by the rule. This sometimes can allow special interest groups to have more of a voice than the public. Special interest groups such as big business or labor can concentrate their influence in the rule-making process. The administrator must remain cognizant of the possibility that special interest groups may want to advance their own agendas over what would be best for the public.

In theory, public hearings are held to solicit the input of all concerned parties. The administrator and staff must play the role of advocate for all those affected by the rule. "Again, it comes back to the idea of fairness, which underlies the legal concept of due process" (Garvey, 1996, p. 140).

Politics of Administrative Choice

It is not uncommon for vote-conscious politicians to be influenced by special interest groups or individuals. At the local level, politicians are more likely to be swayed by influential business leaders, voters, and special interest groups in their election districts.

The bottom line is that politicians are constantly looking for votes. This is why many controversial regulations are delegated to the administrative rule-making process and their administrators. These appointed administrators, it is argued, should be above local politics and will use system-wide planning as the best approach to the problems of a nation of interdependent communities. This method does go a long way toward moderating special interest influence and leveling the playing field, because in most cases the rule-making process is open to public input and scrutiny. However, in the real world, elected officials and special interest groups attempt to influence the outcomes of the administrative process in their favor.

In the fire and emergency services business, there is a slow and steady movement toward progress driven by national standards, mutual and automatic aid agreements, consolidations, and college-level education for fire and emergency services administrators. Professional administrators are helping in the movement away from excessive political influence from special interest groups and elected officials.

NFPA Codes and Standards

The following is an overview of the NFPA's codes and standards (NFPA, 2005):

NFPA's code and standard-making process began in 1896, when a small group of concerned professionals met in Boston to address the inconsistencies in the design and installation of fire sprinkler systems. At that time there were nine different standards for piping size and sprinkler spacing, and these business people realized that unless these discrepancies were resolved, the reliability of these sprinkler systems would be compromised. Working together this small group created a standard for the uniform installation of sprinkler systems, which became the blueprint for NFPA 13, *Standard for the Installation of Sprinkler Systems*.

The NFPA's codes and standards-making process had begun.

For more than 100 years, NFPA has been developing and updating codes and standards concerning all areas of fire safety. An international, non-profit, membership organization, NFPA's mission is to reduce the burden of fire on the quality of life by advocating scientifically based consensus codes and standards, research, and education for fire and related safety issues. While NFPA is involved with extensive fire research and produces numerous fire safety educational programs and materials, its lifeblood is its codes- and standards-making system. Currently there are more than 300 NFPA fire codes and standards used throughout the world.

The NFPA Web page contains several links to information under "Codes and Standards." The electronic version of *NFPA News* contains a list of codes and standards that are in the process of being revised. Also, there is a link to "Drafts of proposed documents" that allows you to review a preliminary document so comments can be prepared and submitted.

The Federal Rule-Making Process

The following is a guide to federal regulations from the U.S. General Printing Office home page:

Federal agencies are required to publish notices of proposed rulemaking in the Federal Register to enable citizens to participate in the decision making process of the Government. This notice and comment procedure is simple. A proposed rule published in the Federal Register notifies the public of a pending regulation. Any person or organization may comment on it directly, either in writing, or orally at a hearing. Many agencies also accept comments via e-mail. The comment period varies, but it usually is 30, 60, or 90 days. For each notice, the Federal Register gives detailed instructions on how, when, and where a viewpoint may be expressed. In addition, agencies must list the name and telephone number of a person to contact for further information. When agencies publish final regulations in the Federal

Register, they must address the significant issues raised in comments and discuss any changes made in response to them. Agencies also may use the notice and comment process to stay in contact with constituents and to solicit their views on various policy and program issues.

Regulations.gov is the U.S. Government web site that makes it easier for you to participate in Federal rulemaking—an essential part of the American democratic process.

On this site, you can find, review, and submit comments on Federal documents that are open for comment and published in the Federal Register, the Government's legal newspaper. As a member of the public, you can submit comments about these regulations, and have the Government take your views into account. (GPO, 2005)

Many national fire and emergency services organizations (and all special interest groups) routinely review the *Federal Register* and circulate any item that may be of interest to the organization. Besides reviewing the information internally, members of these organizations are typically notified of proposed regulations of interest and notices are placed on these organizations' Internet sites. Most of these national organizations have a presence in or near Washington, D.C., so that they can attend the public hearings. Would the following proposal interest anyone in fire safety?

Self-Snuffing Butts—Editorial

When America's largest cigarette maker lobbies Congress to change a law, it's normally a signal to start worrying. Tobacco firms, after all, aren't well-known for seeking out new federal legislation that promotes the public good.

The nation saw an exception recently, though, when Philip Morris, the country's biggest tobacco company, joined health and safety advocates in calling for federal rules to make cigarettes self-extinguishing.

Their stated goal is to reduce the estimated 900 deaths, 2,500 injuries and $400 million in damage caused each year in America by fires started by smoldering cigarettes.

Of course, Philip Morris' goal also may have something to do with the fact that it is the only U.S. company that now manufactures some of the self-snuffing cigarettes.

According to The Associated Press, the firm is concerned about repercussions on its business when New York starts requiring next year [2004] that all cigarettes sold in that state be self-extinguishing. At least four other states are considering similar rules—a potential marketing and distribution nightmare.

"We would have to make different products for different states, theoretically," said Philip Morris lawyer Mark Berlind. "That's the scenario that we hope to avoid."

Fire safety groups have been trying to get a law on self-snuffing cigarettes for years, over the objections of the powerful tobacco lobby.

Cigarette makers didn't support the change in the past because of fears they'd lose business, critics say. Self-extinguishing butts generally burn more slowly, so smokers tend to go through fewer than they otherwise might.

And some smokers reportedly don't like the self-snuffing smokes, complaining that they're inconvenient. "People are calling us and saying, 'Hey, my cigarette is going out and I don't like it,'" Philip Morris spokesman Brendan McCormick told AP.

Such peeves deserve no part in this debate. Smokers are not entitled to faster burning butts when the price for the fires they cause is borne by all of society.

The self-extinguishing cigarette law is long overdue. This is one time—and it may go down as one of the few times in history—when America's biggest cigarette firm is making a good point, despite the fact that it may be doing so for utterly self-serving reasons. (*Arizona Daily Star*, 2002)

Proposals and Comments on Codes, Standards, and Regulations

If the chief or any staff member decides to submit a proposal or comment, it is very important that they research and justify their position thoroughly and accurately. Both the federal government and the NFPA are looking for consensus from the participants, committee members, and public officials. The proposal or comment submitter must be able to convince the regulators or committee members that their proposal has merit. The proposal is more likely to succeed if it is supported by hard data and backed by a well-thought-out line of reasoning.

In most cases, the submitter will not be physically present when the committee or public officials review the proposal or comment. If the submitter can attend a meeting, they will be allowed to present their case personally. Normally, permission to testify must be approved prior to the meeting. Remember, if the submitter cannot attend a meeting, the regulators or committee members will know only what is submitted in writing.

If the submitter has contacts with national organizations, members on committees, or other participants in the process, they should contact these members and voice opinions personally on the subject. They may have more influence in the debate than the submitter does as an individual.

State Regulations

When the federal government adopts a regulation, it may or may not be enforceable by state or local governments, as discussed in Chapter 8. Many fire and emergency services departments that are not compelled to comply with these federal regulations do so because the regulations are incorporated into state laws or union contracts. In addition, volunteer fire fighters must comply in some of these states, but not all.

In one state, volunteers are considered employees when it comes to workman's compensation protection. To gain this coverage they must comply with some minimal requirements, such as a minimum age of 18 years old because firefighting is considered a hazardous occupation.

However, when applying the state's OSHA regulations, volunteers are considered agents or contractors, not employees. Therefore, they do not have to comply with OSHA regulations. However, some of these volunteer organizations contain paid fire fighters who have to comply with the regulations. This situation may be different from state to state.

One OSHA regulation has been adopted by another federal agency that has authority over the non-participating states. The Environmental Protection Agency (EPA) adopted the hazardous materials OSHA regulation 29 *CFR* 1910.120, *Hazardous Waste Operations and Emergency Response*; this action makes it a mandate in all states.

State and Local Building/Fire Codes

Most states and many local governments adopt building and fire codes that regulate the construction and maintenance of structures. There is a lot of variety in how these codes are adopted and enforced. For example, some states have adopted codes that cannot be changed locally, but most allow some local amendments.

At the present time, there are two national organizations promulgating building codes—the NFPA and the International Code Council (ICC). The ICC is a new organization made up of representatives from three model building code groups that have dominated the building codes in this country. Their goal is to produce one building code to be used throughout the United States.

The NFPA has created a building code to complement NFPA standards and codes that previously prescribed only fire and electrical safety features for structures. It has been estimated that up to 50% of the typical building code requirements are fire protection features.

In the near future, many states will have to choose between these two codes for adoption. These regulations can have a major impact on the fire department's ability to protect newer structures. For example, one code may require fire sprinklers and the other may not for a particular building.

A few cities have their own building and fire codes, although there has been a steady trend away from this practice. For example, New York City has decided to adopt the *International Building Code* (IBC, promulgated by the ICC), which will replace its own building code. Commissioner Lancaster (NYC) stated, "As it stands now, our Building Code is the most stringent set of construction regulations in the nation, yet its complexity is seen by many as an impediment to progress. The IBC will allow us to streamline the construction process while not sacrificing the effectiveness of these regulations in keeping our City a safe place to live, work and build" (Lancaster, 2005).

Many cities that have their own building and fire codes reflect the requirements found in one of the national model codes. The specifics in these codes will not be discussed in this book. Building and fire codes are very complex and are best studied through a formal education process such as fire prevention courses at the community college. In many states, inspectors and plans reviewers have to pass a rigorous certification process before they can enforce these codes.

Local Regulations

Generally, few local regulations are adopted to regulate the fire and emergency services. Normally, the regulation is through an arrangement where the chief executive officer (CEO) of the municipality has the authority to supervise the fire and emergency services department and its employees. But, as mentioned previously, it is rare to find a city or county manager or elected official who has expert knowledge about the fire and emergency services profession.

As noted earlier when discussing state regulations, many local regulations are adopted that specify building and fire code requirements. Some have tied these local requirements to the ability of the local fire and emergency services organization to provide public fire protection. Local fire protection requirements that require fire sprinkler protection for structures over a specified height and/or area or that are in a certain use group are common. For example, in Florida all new structures three or more stories in height must be sprinklered, and in Prince George's County, Maryland, all dwellings, including one- and two-family, must be sprinklered.

There are rare exceptions to local regulations of fire and emergency services agencies. For example, a county council passed minimum training and experience requirements for the volunteer fire fighters and officers in its jurisdiction. Because volunteers may not be considered employees at the state level, they are unaffected by the state regulations that require a salaried fire fighter to be trained and certified to professional qualifications. Policies can be implemented at the local level with or without formal legislation.

Zoning Regulations

These regulations are impossible to generalize regarding their origin or intent. There are many unincorporated areas and a few cities, such as Houston, Texas, that have no zoning regulations. In general, in this situation, the local government controls the use of buildings and land to prevent noncompatible uses. For example, it is a bad idea to have heavy commercial development in the middle of an upscale residential neighborhood. Some zoning regulations have their roots in fire prevention and were encouraged by the insurance companies. Some are for esthetics, such as in the city of Washington, D.C., which has a zoning regulation that does not allow any structure to be built higher than the Washington Monument or the U.S. Capitol. This zoning ordinance has an unintended benefit for the fire department—there are no high-rise buildings to contend with.

The best way to gain some understanding of zoning regulations is to review some examples, as follows:

- In residential neighborhoods, small convenience stores can be built only at major intersections.
- For a major commercial development like a covered mall, potential locations will be detailed visually on a map of the municipality and requirements will be specified for adequate transportation to the property via the adjacent road network. The developer may have to pay for road improvements and traffic signals to gain zoning approval from the government.
- In a congested downtown area, a masonry firewall is required on the property line of each building (as a result of a major conflagration in years past).
- If a new building is over 3 miles from a ladder truck, the building has to be protected by fire sprinklers.
- A minimum width of streets was set for new developments such as a gated retirement community.
- In many cases, local governments specify the type of development that can occur in each geographic area by using land use zoning maps.
- Adult entertainment establishments may be limited to commercial zones only, preventing them from using or being close to structures in residential, educational, or professional office areas.
- A comprehensive zoning is approved for a large commercial development and the developer has been encouraged to donate a parcel of property for a new school, library, or fire rescue station.

Again, zoning regulations will not be discussed in any detail in this text. They can be used for many diverse purposes, including fire protection. In general, most are created to control development and establish areas with compatible uses.

Union Contracts

In addition to federal and state regulations, it is common for unionized fire and emergency services employees to participate in a formal negotiating process that ends in a contract between the two parties. Union contracts have become the vehicle to adopt standards and local regulation for fire and emergency services employees. By mutual consent, and with the formal approval of the legislature and executive branch, these documents become legally binding on both parties. In essence, they are virtually the same as a law, regulation, or ordinance.

Unions deserve special attention because they are an important part of the political landscape. In "right to work" states (**FIGURE 9-1**), they will rely on their clout by making contacts and friends with those who are, or have influence on, elected officials. In other states many unions have mandatory bargaining rights. These are very important negotiations because the union's goals can be inconsistent with the municipal and agency administration goals.

The relationship between union members and the administration can be considered adversarial and confrontational at times. Even though one of the administration's goals may be the improvement of services, it could conflict with what the union believes are the rightful privileges of its members.

For example, the administration may want to improve the likelihood that members will be at their sharpest intellect and well-rested when responding to emergencies, and therefore propose to convert to duty shifts of no longer than 8 hours. Several studies have shown that work hours in excess of 8 per shift have an adverse effect on a worker's mental sharpness. This may be especially critical for those members who are paramedics. Theoretically, this proposal would have the effect of providing a better emergency medical provider who would make fewer mistakes in providing medication and treatment for patients.

However, the union, by virtue of its primary duty, would probably oppose this change because it would be a hardship on many of its members. It is common for fire fighters who work 24-hour shifts to have a

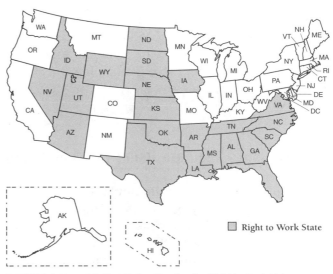

FIGURE 9-1 Map of United States Showing the Right to Work States

second job, and in many cases they actually spend almost as many hours at the part-time job as at the primary job.

Remember, the main goals of the union are to maximize pay and benefits while minimizing management's rights to make changes. The fire and emergency services administrator strives to implement constructive change at the lowest practical cost to the taxpayers. These different goals often create conflicts.

The administrator must handle the following important concepts that may help to facilitate change:

- Determine which decisions and policies are required by law or ordinance to be negotiated with the union.
- Those policies that are not required by law or ordinance can be delegated to a nonbinding task group made up of representatives of all interested parties.
- Understand the language in the existing labor contract, especially the concept of *past practice*.

If a decision is not a mandatory subject of bargaining, the chief does not have to secure the union's permission before implementation. The bargaining process only requires good faith discussion with the union and a willingness to include the union's ideas and concerns in the final outcome. The union should be given a full opportunity to express all its viewpoints.

However, in some jurisdictions a formal impasse process may be present. If both parties cannot come to agreement, an arbitrator can be brought in to hear and decide the issue. This is called binding arbitration and is generally preferred by unions. Without this type of an outside judge, the administration has the advantage because they do not have to agree to anything, even a reasonable compromise.

Even if mandatory negotiations are not used in the jurisdiction, the process of having formal meetings between the administration and members is always a good idea. It is beneficial to the members' morale and useful in implementing change, helping the organization to function more smoothly and providing better service to the public.

Governments Regulate Using Taxes and Fees

If the federal government did not subsidize the purchase of single-family dwellings, many more people would be living in apartments in this country. Through the tax code, the federal government allows taxpayers to keep more of their own money by reducing federal taxes. They do this by allowing the tax-payer to deduct interest and real estate taxes from their income before computing federal taxes each year.

This is a very powerful governmental authority. Financially, they can convince taxpayers to change their decisions about a purchase or where to obtain services by providing economic incentives. This can also work in a negative fashion, such as when the government taxes something like cigarettes to raise their price and discourage consumption.

After a number of failed attempts to gain approval for a mandatory single-family dwelling sprinkler ordinance, one county adopted a reduction in local real estate taxes for those new single-family dwellings that had fire sprinklers installed. This reduction was for the first few years and would offset most of the increase in the mortgage payment needed to pay for the sprinkler installation.

When a sprinkler system is installed in a new home, the cost is reflected in the total cost. The total sales price is used to set the real estate tax. One could ask the following questions: Why should someone who installs a fire safety device that reduces the demand for public fire protection for the home have to pay a higher tax than someone who does not install this life-saving feature? This is a benefit to the community at large, so why should it not be reflected in lower taxes?

At the local and state level, it is common to find tax incentives meant to encourage companies to relocate into their cities or states. Generally, the real estate tax is reduced or eliminated for the first few years. This allows the company to offset most of the cost for moving to a new location. Eventually, the taxes return to normal levels.

At the federal level, Congressman Curt Weldon of Pennsylvania has introduced a bill (HR 1131, The Fire Sprinkler Incentive Act) that would allow an accelerated depreciation, from 39 years to 5, for the cost of automatic sprinkler installations on the federal income tax. This has the effect of providing a fiscal incentive for the owner, who will experience lower federal taxes as a result of this deduction. Section 2, Findings, in HR 1131-2005 has an excellent overview of justifications for automatic sprinklers. In March 2005, this bill was sent to committee and has 118 sponsors. A companion bill in the Senate (S.512) has 14 sponsors. This proposed legislation covers only existing buildings.

Again, this is a very complex area where the tax laws can be used by government to regulate people and development. Although rare, they can be used to create incentives for the installation of fire protection systems.

Fire Service Laws and Regulations

Many states do not have laws or regulations that require fire fighters to be trained and certified to minimum levels of professional qualifications. However in Florida, for example, a 480-hour fire-fighter training program must be completed to qualify to take a state exam that includes a skills test and a written exam to prove competency at the NFPA Fire Fighter II level. If this exam is completed successfully, the applicant is eligible to be employed as a fire fighter in that state. Volunteers are required to achieve the Fire Fighter I state certification. In some states, volunteers are exempt from the state laws, but there may be a county ordinance, municipal policy, or volunteer company rule that requires volunteer fire fighters to be trained and certified. In states that do not have minimum standards legislation for fire fighters, either paid or volunteer, each agency, city, or county can have its own requirements.

In addition, there are differences in the quantity and quality of some of these training and certification programs throughout the country. Although they all follow the same NFPA standards for professional qualifications, some seem deficient in content and testing. At the present time there does not seem to be consensus among fire service leaders in this area of standardization.

The standards used for EMS are mandated by the Department of Transportation and specify the minimum number of hours needed to master a certification level such as EMT-Basic, which is 120 hours. This is a national standard that contains both the content and minimum training time for competency. This helps assure that EMS training and certification programs throughout the country are equivalent to each other.

OSHA regulations are of great interest to the fire service, even in states without a *state plan*. Again, most of these regulations do not apply to non-state plan OSHA states unless noted.

OSHA 29 CFR 1910.146: Permit-Required Confined Spaces

This federal regulation covers employees who enter areas with limited or restricted means for entry and exit, such as trenches, pipes, tanks, and vaults—all are areas with atmospheres that can be oxygen deficient and potentially trap a person.

This regulation requires training, proper equipment and written procedures for all employees operating in these hazardous areas. When the fire department re-

ceives a 9-1-1 call that someone has been overcome and is unconscious in a confined space, the fire fighters must enter the confined space to rescue the victim.

Several fire fighters have died when they entered these atmospheres without self-contained breathing apparatus (SCBA). They were overcome very quickly and could not be rescued by others because they also were not prepared to enter the immediately dangerous to life and health (IDLH) area.

OSHA 29 CFR 1910.134: Respiratory Protection

Among other requirements for training and equipment, this OSHA regulation requires that four fire fighters be on the scene before an interior structural fire can be attacked. The fire department will generally issue a standard operating procedure (SOP) that requires the assembly of four fire fighters before entry into a structural fire (IDLH) can be attempted. Many municipalities are increasing their staffing on engines to four fire fighters to comply with this federal regulation. If a department staffs its engine companies with fewer than four persons, in the case of a working structural fire, they will have to wait until another company (or volunteer/call fire fighters) arrives to make up the four fire fighters. In these situations, the best opportunity to start extinguishment would be using an outside attack. This can be successful if the room where the majority of the fire is located has a window or door to the outside. This is frequently the case in residential occupancies. For the safety of any occupants who remain inside, the attack should be done with a solid stream, which tends not to push the fire, heat, and smoke into other areas of the structures where victims may still be present. This is a tactic that is practiced infrequently and is frowned upon by some individuals in the fire profession. This outside fire attack should be part of the training and SOPs if there are fewer than four fire fighters on each engine.

The following are excerpts from a document prepared by the International Association of Fire Chiefs (IAFC) and the International Association of Fire Fighters (IAFF) on this regulation, typically abbreviated as *two in/two out*:

- OSHA defines structures that are involved in fire beyond the incipient stage as IDLH atmospheres. In these atmospheres, OSHA requires that personnel use self-contained breathing apparatus (SCBA), that a minimum of two fire fighters work as a team inside the structure, and that a minimum of two fire fighters be on standby outside the structure to provide assistance or perform rescue.

- Fire fighters operating in the interior of the structure must operate in a buddy system and maintain voice or visual contact with one another at all times. This assists in assuring accountability within the team.
- Any task that the outside fire fighter(s) performs while in standby rescue status must not interfere with the responsibility to account for those individuals in the hazard area. Any task, evolution, duty, or function being performed by the standby individual(s) must be such that the work can be abandoned, without placing any employee at additional risk, if rescue or other assistance is needed.
- OSHA regulations recognize deviations to regulations in an emergency operation where immediate action is necessary to save a life. However, such deviations from the regulations must be exceptions and not de facto standard practices.
- Fire departments must develop and implement standard operating procedures addressing fire ground operations and the two-in/two-out procedures to demonstrate compliance. Fire department training programs must ensure that fire fighters understand and implement appropriate two-in/two-out procedures.
- It is unfortunate that all U.S. and Canadian fire fighters are not covered by the OSHA respiratory protection standard. However, the two-in/two-out requirements are the minimum acceptable standard for safe fire ground operations for all fire fighters when self-contained breathing apparatus is used. (IAFF & IAFC, 1998)

Finally, there are requirements for fit testing and medical evaluation. The medical evaluation consists of the user answering a number of questions relating to medical and physical fitness. If necessary, a medical examination by a physician may be required depending on the answers.

OSHA 29 CFR 1910.1030: Occupational Exposure to Bloodborne Pathogens

Briefly, this rule requires the employer to develop a plan and take action that will reduce or eliminate the exposure to several diseases that are transmitted by body fluids. These requirements are typically covered in a fire and emergency services agency SOP detailing response to emergency medical calls.

OSHA 29 CFR 1910.120: Hazardous Waste Operations and Emergency Response (HAZWOPER)

This regulation was also adopted by the EPA, so it covers all persons who would have contact with hazardous materials in all states including those not covered by OSHA state plans. Therefore, all fire fighters and other emergency responders such as police and EMS personnel who respond to hazardous materials incidents must be trained to a minimum level of competency.

For fire fighters, this is the Operations level. Those who have a need to protect themselves and not take any emergency action, such as police officers and EMS support staff, are required to meet the Awareness level, which is the lowest level of training and competency.

This OSHA regulation was based on a 1986 draft of NFPA 472, *Standard for Professional Competence of Responders to Hazardous Materials Incidents*. It is generally a good idea to base training and certification on the latest edition of NFPA standards. OSHA has not updated its hazardous materials regulation since 1989; however, the NFPA has updated its standard three times. OSHA recognizes and accepts more recent standards as equivalent to the original rule. To gain a better understanding of this regulation and complying with its mandates, the *NFPA Hazardous Materials Response Handbook* provides an in-depth explanation of this standard and several useful appendix articles dealing with issues of organizing a hazardous materials response team.

Typically, basic hazardous materials training (Awareness and Operation levels) is incorporated into recruit fire-fighter training programs. These are minimum federal requirements for responders to hazardous materials incidents.

OSHA 29 CFR 1910.156: Fire Brigades

This OSHA rule, first adopted in 1980, has been largely ignored by the public fire service. Many of its requirements are nonspecific and therefore hard to understand. Also, the title *Fire Brigades* seems to convince readers that it is not applicable to public fire departments. Fire brigade is another name for fire department, but is more common in other countries and in industry. Items required by this rule are:

- An organizational statement
- Training and education similar to fire training in states such as Maryland, Georgia, and Washington

- In-service training
- Protective clothing and respiratory protection at no cost to employees

In the appendix, it suggests:

- Preplanning
- A physical fitness program
- Fire officer training and education

In general, most of these same items can be found in NFPA 1500, *Standard for Fire Department Occupational Safety and Health Program*. The NFPA standard has specific language regarding requirements for safety policies, procedures, and equipment, and therefore it is easier to enforce and to comply with its requirements. The *Fire Brigades* regulation can be used in conjunction with the *General Duty Clause*, as explained in the following section.

OSHA's General Duty Clause

This clause requires each employer to furnish a place of employment safe from hazards that are likely to cause injury, sickness, or death. OSHA and court precedent have shown that in the absence of specific or up-to-date OSHA regulations, a national consensus standard such as NFPA 1500 can be used to assess compliance. OSHA and "state plan" states have issued numerous violation notices for failure to comply with nationally recognized safety standards.

For example, fire fighters in a large suburban county filed a complaint (many years ago) with their state OSHA agency to stop the practice of fire fighters riding on the back step of fire engines. Even though there is no specific requirement in the federal OSHA *Fire Brigades* regulation, the state agency cited the *General Duty Clause* along with NFPA 1500 when they filed a violation against the county. The practice was halted immediately and the proposed fine was eliminated.

It is very important to keep this expansive safety clause at the top of the to-do list. As a minimum, the department should consider complete compliance with NFPA 1500 for the fire and emergency services agency. If not, the chief and the organization could be open to civil liability and/or state OSHA citations.

NFPA Standards

At the top of the list of NFPA standards that have had a substantial effect on the fire service are 1500, *Standard on Fire Department Occupational Safety and Health Program*, and 1710, *Standard for the Organization and Deployment of Fire Suppression Operations, Emergency Medical Operations, and Special Operations to the Public by Career Fire Departments*. The first edi-

tion of NFPA 1710 was adopted in 2001. NFPA 1500 has been around since 1987, when it was first issued as a standard.

At that time, there were many who proclaimed that NFPA 1500 would be too expensive and impractical to realistically implement. It would be the end of many smaller departments, who would collapse under the substantial economic burden. Although most fire departments have not complied with the standard 100%, many have made substantial changes to provide a safer work environment for fire fighters. As with any change, they are typically evolutionary rather than revolutionary.

NFPA 1500 can also be looked at as an umbrella document that adopts by reference many other NFPA standards such as NFPA 1001, *Standard for Fire Fighter Professional Qualifications*; NFPA 1021, *Standard for Fire Officer Professional Qualifications*; NFPA 1403, *Standard on Live Fire Training Evolutions*; NFPA 1901, *Standards for Automotive Fire Apparatus*; NFPA 1971, *Standard on Protective Ensemble for Structural Fire Fighting*; and NFPA 1981, *Standard on Open-Circuit Self-Contained Breathing Apparatus for Fire and Emergency Services*. These standards, in addition to providing for fire fighters' safety, also provide fire fighters with the training, tools, and protective clothing needed to operate effectively and professionally in mitigating dangerous uncontrolled fires and other emergencies that threaten to destroy property and can injure or kill occupants.

One aspect of this standard that still has not been defined clearly is physical fitness. Although everyone would agree that firefighting requires high levels of strength and endurance, the NFPA committee assigned the responsibility for defining this standard (Fire Service Occupational Safety and Health committee) has not been able to come to a consensus on exact levels.

NFPA 1710, *Standard for the Organization and Deployment of Fire Suppression Operations, Emergency Medical Operations, and Special Operations to the Public by Career Fire Departments*

This document will probably have more impact on the fire and emergency services than any other except NFPA 1500. The word *career*, as used in NFPA 1710, does not necessarily mean the departments are fully paid. *Career fire departments* include departments that rely on fire fighters who are "on-duty" in the fire station. Part or fully volunteer organizations that schedule their fire fighters to cover shifts can use this standard.

This discussion is based in part on *NFPA 1710: A Decision Guide*, published by the IAFC and IAFF (2002).

One of the first items that needs to be clarified is the issue of which fire departments fall into the category of "substantially all career fire departments." For completely paid departments this is not an issue, but for many combination departments this may not be easy to determine. The following are several examples that can be used for guidance:

- A department has 20 volunteer members and 12 paid fire fighters. This seems like an easy call and this department would *not* fall under NFPA 1710 (i.e., greater than 50% career). The run reports listing fire fighters actually riding the apparatus, when totaled over a year, show an average of 2.5 volunteers and 3 paid personnel per call. The paid fire fighters work a 56-hour shift with four fire fighters assigned per shift and a minimum staffing of three. This one example shows why it is important to know who is actually riding the fire apparatus, not how many volunteer members or total paid employees are reported.
- Who takes the first attack line into the structure?
- How many calls are staffed by career/volunteer workers?
- Are the volunteers receiving more than nominal compensation (paid on call)? In some towns, an employee of the town or employees of the fire company, such as a dispatcher, maintenance person, or the like, is allowed to respond to emergency calls with the town's volunteer fire department. These people are actually on duty and should be counted as career members. For example, in Florida, employees of a town who respond in this fashion are consider paid fire fighters and must comply with the state's fire-fighter certification requirements.
- The following is the NFPA 1710 appendix definition for *member*: "A fire department member can be a full-time or part-time employee or a paid or *unpaid volunteer*, can occupy any position or rank within the fire department, and can engage in emergency operations" (NFPA, 2001).
- If a volunteer fire fighter is in the station when the call comes in, he or she is "on-duty" and can be counted towards the "sub-

stantially all career." By definition, a volunteer can be included as a career member in NFPA 1710. This also allows volunteers who are on-duty in the station to count towards the minimum number of fire fighters on fire companies and the total first alarm assignment. It is theoretically possible to comply with this standard with an all-volunteer company that assigns their members to specific duty times.

There are many legal implications of not complying with this standard. However, even if the department's liability exposure is small, it is still a good idea to use the NFPA standard as a benchmark for planning for public fire protection in the community. The community and its citizens deserve to know what level of public fire protection they are receiving. This is truly a landmark planning document for the fire and emergency services.

NFPA 1710 Response Times and Staffing The IAFF and IAFC document does a good job detailing the requirements for response times and staffing. The document suggests that there are a number of ways to arrange the response of fire fighters to the scene. The document lists three specific circumstances for organizing as a "multi-piece company":

1. The use of a fire department personnel's vehicle, such as a car or van, if the main fire apparatus does not have adequate seating
2. An engine and a water tanker
3. An engine and an EMS unit (ambulance or rescue)

Four-Person Companies The superior method of complying with this standard is to provide on-duty, in-station staffing of three fire fighters and one officer per company. The NFPA 1710 document contains many references to studies that support this statement, and chiefs and elected/appointed officials are encouraged to review them if they do not understand or have a different opinion (see IAFC & IAFF, 2002, Section 5, Resources). One of the main reasons why this method works best is that each fire fighter is prepared to work as part of a company and will be ready to perform an assigned role. Fire fighters arriving one at a time will have to fill a particular position that is unknown until they arrive on the emergency scene. Even with well-trained fire fighters, this can result in an uncoordinated effort.

The on-duty company in the station can be analogous to a football team's huddle where the quarterback calls the play and the players are assigned positions ahead of the snap of the ball. Try to imagine a referee blowing a whistle, and then 11 players running onto the field and assuming any position open. Once all the positions were filled, whoever became the quarterback would then call the play. How successful would this team be against others that operated from a huddle?

One staffing method for a company could be to assign four fire fighters, with two being paramedics. They would operate an engine and paramedic unit combination. To make this cost effective, appropriate charges for emergency medical transports would help to offset the salaries of the two EMS members. The one drawback to this method can be seen when the paramedic unit is transporting a patient to the hospital, or returning, and a fire call is received. At this time the first due company will arrive with only two fire fighters, necessitating waiting until others arrive to start an interior attack. This works well until the call volume exceeds the point that affects the company's ability to provide a 4-minute response with 90% reliability. Then separate staffing for the paramedic unit would be needed.

Response Time Requirements The response time requirements for a structural fire call state that the first due engine must arrive within 4 minutes (one additional minute is allowed for turnout) and/or the full first alarm assignment must be made within 8 minutes (and again, one minute is allowed for turnout). For record-keeping purposes, a successful response would be counted when either of these is achieved. This was done with the understanding that engine companies may be out of position, such as during preincident planning, fire prevention education, physical fitness, or training exercises, when notified to respond. It is not uncommon for engine companies to be out of the station but in their first due areas.

The 90% Level All of the response times are measured at the 90% level. This allows for exceptions due to the nature of a situation. For example, when a company is responding to or working at an emergency incident, another call in its first due area will be answered by a company coming from a distant station. It would be nearly impossible and unreasonably expensive to provide anything close to full compliance.

The number of times that a company will be busy when a second call comes in is in direct proportion to call volumes. In other words, the greater the number of calls per year, the greater the busy time and percentage of calls not meeting the maximum response time standard. Although there are many influential circumstances, generally a call volume in the range of 3,000 annually will create a situation where the response time will frequently not be met. The ISO *Municipal Fire Grading Schedule* once contained the requirement that a second company would be needed if more than 2,500 calls were responded to in a year.

In some cases there are policy solutions for these excessive numbers of responses. The following are a few examples:

- Dispatch policy can sometimes create additional responses to alarms that may not be necessary. For example, some departments send a full first alarm assignment to an automatic alarm whereas others send one engine.
- A vehicle crash may generate the automatic response of a paramedic unit, an engine, and a heavy rescue squad in one jurisdiction, and in another only the paramedic unit responds, unless additional information indicates a serious accident with occupants trapped.
- Other than the required companies that must be dispatched on a structural fire (totaling 14 fire fighters), review all dispatches that send more than one unit and calculate the percentage used for units other than the closest.
- A second unit in the same station will cut the call volume substantially. To make this work effectively, only one unit should be dispatched on any fire call, leaving the other unit (engine) to answer a second call in the first due area within the expected 4 minutes.

One final word about NFPA 1710: This is truly an innovative document and will have a distinct impact on fire and EMS services in the future. It is a planning document that will point out weaknesses in fire protection and emergency medical service that are quantifiable. This will bring the discussion of funding these weaknesses into focus so that citizens and both elected and appointed officials have a good understanding of where their communities have adequate fire and EMS coverage and areas that are beyond response time standards. In some jurisdictions with areas that are undeveloped or that have low population densities such as farmlands, the 4-minute response time would not be justified.

Regulations Dealing with Hiring and Personnel Issues

There are numerous rules, regulations, and laws that pertain to hiring and personnel actions for employees, such as the Fair Labor Standards Act (FLSA), Equal Employment Opportunity Commission (EEOC), Americans with Disabilities Act (ADA), affirmative action, discrimination, age issues, and the Family and Medical Leave Act. These were previously discussed in Chapter 5, Human Resources Management.

The National Highway Traffic Safety Administration (NHTSA)

The National Highway Traffic Safety Administration was created in 1970 to carry out safety programs aimed at reducing the number of vehicle crashes and the subsequent injuries and deaths. It either directly or indirectly initiated data collection, investigated safety defects, reduced the threat of drunk drivers, promoted the use of safety belts and airbags, and provided consumer information on motor vehicle safety topics. It has also been instrumental in implementing a nationwide 9-1-1 emergency reporting system and determining the training for emergency medical competencies for responders who provide treatment to injured drivers and passengers.

From the use of medivac helicopters to shock trauma centers, NHTSA has been directly involved (**FIGURE 9-2**). It has also been responsible for the standards for training and certification from EMT-Basic to Paramedics. This was all done to increase the likelihood of survival after a motor vehicle crash.

Legal and Court Issues

Legal issues and opinions are in a continual state of change as a result of new legislation and legal decisions from numerous judges throughout the country.

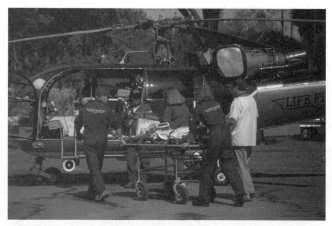

FIGURE 9-2 Medivac Helicopter

This section provides a snapshot of legal opinions up to 2005.

Keep in mind a couple of points when reviewing this section and articles from periodicals. The first is the hierarchy of the judicial system in this country. When reviewing decisions from a court, unless the Supreme Court of the United States has made a determination about a legal issue, the issue can always be appealed for another opinion. For example, a lawyer may explain that a circuit court awarded a large settlement to an employee as a result of some personal action by a fire and emergency services organization. If the administration is using this information to determine if the department's policy is illegal or not defensible in court, this information should have less of an influence on actions than a decision from a higher appeals court. This is especially true if the information pertains to an out of court settlement.

Second, each lawyer, attorney, or counselor may have a different opinion about a particular law or regulation. Just as a patient should do if their doctor recommends major surgery, chiefs would be smart to get a second opinion and do some research on their own.

A couple of things can be done to protect the administration and the organization from legal problems. Where available, follow the requirements of an appropriate professional standard or regulation (e.g., NFPA or OSHA). Ensure that all the department's policies treat all people equally. Demand that any legal advice be documented by case histories from courts at the appeals level or above. Ask the lawyer to present all sides of the argument, not just the legal precedent that supports their opinion or the department's policy.

The Court System

The better the administration understands the legal process and the court system, the better it will be able to protect the interests of the fire and emergency services organization and the public it serves. Remember, if the department is persuaded by legal advice to reduce professional standards for its members, this will be compromising a critical emergency service to the public. This is a very serious and grave responsibility for the administration.

For example, after discussions with an attorney, it would be fairly easy to decide not to go to court when faced with a job applicant's threat of a lawsuit. "Beware of the attorney who wants to settle too quickly at the fire department's expense. These actions may set unwarranted precedents with the department" (Edwards, 2000, p. 53).

Federal and state courts have three separate levels: the district court, the appellate court, and the Supreme Court. In most cases, federal law takes precedence over state law except for laws and regulations that fall under the protection of states' rights in the 11th Amendment to the Constitution of the United States. At the district court level, rulings are not considered precedent setting. Only the decision of the appeals courts is considered precedent setting.

When appealed, the issue always revolves around an interpretation of law or the conduct of the district court. Decisions from these courts are precedent setting, but more strongly so in the individual circuit. It is possible for appeals courts to make decisions that are in conflict with other appeals courts in other districts. That is why the Supreme Court has the final authority, and its decisions become precedent for the entire nation, or the entire state if it is a state supreme court. It is also significant when the Supreme Court refuses to hear a case, which is its prerogative, indicating that it supports the lower court's decision.

Attorneys

"One of the most common errors of fire service personnel managers is not seeking legal advice early enough in the decision-making process" (Edwards, 2000, p. 51). Another mistake is getting advice from an attorney who may not have the appropriate background or best interests of the fire and emergency services organization in mind.

Attorneys are similar to anyone else who must help the agency in that they may not understand the critical internal workings of an emergency organization. Educate them, starting with the basics. They may focus on the legalities of a situation that has extenuating circumstances. For example, the fire and emergency services organization may have a strict policy that any new applicant may not have anything in their background indicating that they cannot be trusted, such as a criminal record, excessive driving citations, or poor credit. You would need to explain to the attorney that there are many occasions when members will be in private homes during an emergency call, and the organization cannot tolerate any behavior, for example, leading to stealing private property or any other criminal activity.

However, sometimes the attorney may attempt to make the policy decision under the guise of legal advice. . . . Make sure that you understand the difference between legal advice and policy advice. In my experience often what occurs is that the attorney is focusing on the legalities but does not understand the policy ramifications because the fire department has not taken the time to properly brief the attorney of its situation. (Edwards, 2000, p. 52)

In some situations it would be wise to ask for another opinion and do some independent research. Contact peers and check with professional organizations. Look for articles that may have been written about the subject. Do not rely on one source. Many individuals have their own agendas.

If the case is lost at the district court level, make sure the department's attorney and the elected officials will support an appeal. On any given day, a judge can give a decision that is based not on the law but on their own personal feelings about the law or the specific circumstances. Do not compromise when the issue is very important to the organization and its ability to provide quality service to the public. Always be prepared to appeal any misguided or inaccurate ruling. Make sure the other side is aware of the department's resolve. This can sometimes keep the department and the chief out of court.

If the organization or the municipality is too small to have a staff attorney, or if the attorney does not have enough time to provide good service, ask if the department can select an outside attorney. This is especially helpful if the loss of the case will have a serious adverse impact on the organization. If the present attorney does not seem to be meeting the department's needs, or as soon as the administration recognizes the inadequacy of either appropriate time or expertise or a lack of commitment to winning the case, insist on changing attorneys.

Legal Aspects of the Fire and Emergency Services

The following is a list of subjects and items with which fire and emergency services administrators should be familiar:

- All legal risks, both criminal and civil
- Human resources and fiscal management
- Fire prevention ordinances
- Administrative law
- Common law, statutory law, administrative rules, constitutional law, and court decisions
- Criminal vs. civil law

- The U.S. Supreme Court has repeatedly held that summary governmental action taken in emergencies and designed to protect the public health, safety, and general welfare does not violate the due process clause of the Constitution and is appropriate. Governments and their agents are given great leeway in adopting and implementing summary actions under their police powers.
- Failure to train properly or not complying with a "standard of care," such as professionally recognized standards, could give rise to liability.
- *Fireman's rule:* A fire fighter cannot sue a negligent person for injuries from the incident that created the need for the fire fighter's presence at the scene.
- *Sovereign immunity:* In its purest sense, a government or its employees cannot be sued no matter how negligent or egregious their behavior was. In recent times, state legislatures and the courts have been eroding and sometimes abolishing sovereign immunity.

Checklist on Legal Issues

- First and foremost, don't become paralyzed by fear of liability.
- Get the facts. Do independent research.
- Check with more than one attorney.
- Check on the department's immunity.
- Satisfy and comply with all laws, regulations, and safety standards.
- Perform all duties imposed by law or regulation.
- Be fair.
- Train, train, and train.
- Demand professional performance (i.e., to NFPA standards).
- Consult recognized experts (remember, not just one).
- Have access to a well-informed attorney.
- Document all actions in writing.
- Be proactive in proposing changes or new laws and regulations when necessary.
- Take prompt action on complaints.
- Follow up on discipline and violations of department SOPs.
- Know the professional expectations of the job (experience, training, and education).
- Know the statutes, regulations, and court cases.
- Review legal and occupational publications.

- Treat all employees in the same, fair, even-handed way.
- Finally, keep customer service first when creating the department's goals.

Discussion Questions

This assignment is to complete the NFPA 1500 Worksheet found in Appendix B of the standard. It is not necessary to complete all columns; just the columns marked "Compliance, Partial Compliance and Compliance with Administrative Action."

1. First, contact an official representative of a department that will help provide this information for the assignment. After they have agreed to participate, send them a written note thanking them, confirming a meeting date and time, and advising them of the need for written documentation for the sections listed. Send them a copy of the worksheet form with the sections highlighted for substantiation. Finally, advise them in writing that this information will not be released to anyone other than faculty.

    Remember, for an accurate understanding of a section or paragraph, you may need to do some research, starting with reviewing the NFPA handbook. If you are not convinced that a requirement has been explained accurately, contact another knowledgeable person for an opinion. An e-mail to an NFPA staff person is always a good option.

    Provide specific details and documentation of the following: Sections 4.4.4, 4.5, 4.8, and 4.104; Chapter 5; Section 6.2; Chapter 8; and Sections 10.1, 10.2, and 10.3. For these sections, provide copies of written official documents including SOPs and policy directives supporting compliance. In addition, provide a short biography of the person who helped complete the NFPA 1500 worksheet.

2. Find a recent article from a fire, EMS, or emergency services periodical that discusses a recent appeals or Supreme Court decision. Summarize the article.

    Now find two more articles that discuss the same issue either before or after the court's decision. Compare and contrast the articles. Once you think you have a good understanding of the issues and the court decision, find an attorney who will give some advice. Explain that it is a school project. Many attorneys will give a student a limited amount of time without professional charges. If you are a fire fighter, check with a city, county, or state attorney.

Finally, after writing down a synopsis of the above, use one section to express personal thoughts about the case. Limit the answer to no more than two pages.

References

Arizona Daily Star. Self-snuffing butts. *Arizona Daily Star,* April 29, 2002, p. B4.

Edwards, Steven T. *Fire Service Personnel Management.* Upper Saddle River, NJ: Brady/Prentice Hall Health, 2000.

Garvey, Gerald. *Public Administration: The Profession and the Practice.* New York: St. Martin's Press, 1996.

GPO (U.S. Government Printing Office). *Your One Stop Site to Comment on Federal Regulations.* http://www.regulations.gov (accessed July 24, 2005), 2005.

IAFC & IAFF. *NFPA 1710: A Decision Guide.* Fairfax, VA: IAFC, 2002.

IAFF & IAFC. *IAFF/IAFC Two in/Two out Questions and Answers.* http://www.iaff.org/safe/pdfs/2in2out.pdf (accessed July 24, 2005), 1998.

Johnson, Robert. Lines of duty: As blazes get fewer, firefighters take on new emergency roles—Cities complain about costs, but safety group's plans may help fuel the trend. *Wall Street Journal.* February 7, 2001. p. 1.

Lancaster, P.J. *Press release.* New York City Department of Buildings. http://www.nyc.gov/html/dob/html/news/intro_478a_testimony.shtml (accessed July 24, 2005), June 24, 2005.

NFPA. *Fire Department Occupational Health and Safety Standards Handbook.* Quincy, MA: NFPA, 1997.

NFPA. *How the Code Process Works.* http://www.nfpa.org/itemDetail.asp?categoryID=162&itemID=17174&URL=Codes%20and%20Standards/Code%20development%20process/How%20the%20code%20process%20works (accessed June 17, 2003 and September 29, 2005), 2002, 2005.

NFPA. *Resources for Fire Department Occupational Safety and Health.* Quincy, MA: NFPA, 2003.

NFPA. *1710, Standard for the Organization and Deployment of Fire Supression Operations, Emergency Medical Operations, and Special Operations to the Public by Career Fire Departments.* Quincy, MA: NFPA, 2001.

OMB. *Circular No. A-119, Federal Participation in the Development and Use of Voluntary; Consensus Standards and in Conformity Assessment Activities.* http://www.whitehouse.gov/ omb/circulars/a119/a119.html (accessed July 24, 2005). February 10, 1998.

OSHA. *December OSHA Facts.* http://www.osha.gov/as/opa/oshafacts.html (accessed June 16, 2003), 2004.

10

Ethics

Knowledge Objectives

- Comprehend the creation, effect, and importance of a professional organization's code of ethics.
- Understand the impact and influence that lying and moral choices have on fire and emergency services administration.
- Examine and comprehend the effect that ethics has on cost-benefit analysis in the fire and emergency services.
- Identify the fiscal responsibility and accountability that the administration has to the public.
- Examine the role of the administrator as a caretaker of public trust.
- Comprehend and understand the moral and ethical justifications for government oversight of the public fire and emergency services.

Ethical Behavior

The *American Heritage Dictionary* defines ethics as "A principal of right or good conduct; or a system of moral principles of values" (*AHD*, 2005). Ethical behavior is behavior based on a system of moral principles. Although it can seem like a straightforward method of determining appropriate decisions, it can be very difficult to judge each action in the context of increasingly complex situations and ever-changing policies. Some would deem only those actions that are *illegal*, such as robbing a bank, as being wrong. But illegal actions are only a subset of the larger category know as lying.

In administering a fire and emergency services agency, many actions and decisions have to be made daily. Each has the potential of having ethical considerations and consequences. Because many of these decisions are ultimately made by the chief, the obvious question to ask is: Is the chief doing the *morally right*

actions in all administrative functions? Not an easy question to answer.

How does a chief know when his co-worker or boss is lying? Are there situations when it is morally right to tell a lie or mislead a person or persons so that they have an incorrect understanding of a situation or idea?

Justification

Imagine a person in their kitchen at home when two very young children come bounding through the door. They tell the adult that a mad person is chasing them with a knife. They are quickly directed to hide in a closet. Almost immediately after the children close the closet door, a man bursts through the kitchen door with a large machete in his hand. He asks the adult in the kitchen if he has seen two children. The adult lies to him and tells him he has not seen any children. The intruder goes away.

Is this person justified in lying? The rational answer would be yes.

Trying to determine if something is a lie can be very baffling (**FIGURE 10-1**); the following excerpts from *Lying—Moral Choice in Public and Private Life* (1999) by Sissela Bok help illustrate the complexity of this issue:

- "Convinced that they know the truth—whether in religion or politics—enthusiasts may regard lies for the sake of this truth as justifiable. They may perpetrate so-called pious frauds to convert the unbelieving or strengthen the conviction of the faithful. They see nothing wrong in telling untruths for what they regard as a much 'higher' truth" (Bok, 1999, p. 7).
- "I shall define as a lie any intentionally deceptive message which is stated." "Deception, then, is the larger category, and lying forms part of it." "Thus, if you are asked whether you broke somebody's vase, you could answer 'No,' adding in your own mind the mental reservation 'not last year' to make the statement a true one" (Bok, 1999, pp. 13–14).
- "Paradoxically, once his word [the person lying] is no longer trusted, he will be left with greatly decreased power—even though a lie often does bring at least a short-term gain in power over those deceived" (Bok, 1999, p. 26).
- "Deceit [lying] and violence—these are the two forms of deliberate assault on human beings. Both can coerce people into acting against their will" (Bok, 1999, p. 18).

"Let's get one thing straight. I don't want your money, I want your respect."

FIGURE 10-1 Is This Person Lying? *Source:* ©www.cartoonstock.com

- "Those who learn that they have been lied to in an important matter—say, the identity of their parents, the affection of their spouse, or the integrity of their government—are resentful, disappointed, and suspicious. They feel wronged" (Bok, 1999, p. 20).
- "What if all government officials felt similarly free to deceive provided they believed the deception genuinely necessary to achieve some important public end? The trouble is that those who make such calculations are always susceptible to bias." "When political representatives or entire governments arrogate to themselves the right to lie, they take power from the public that would not have been given up voluntarily" (Bok, 1999, p. 173).

A common situation in the fire service is the response of a fire engine with one, two, or three fire fighters. When the general public sees the fire engine responding with its red lights and siren operating, they are unable to see how many fire fighters are on board. Even though the minimum number of fire fighters on the truck has been set by a national standard, the general public has no knowledge of how many fire fighters are needed to do a safe and effective job. The fire and emergency services administrator can easily mislead the public and elected officials into thinking that whatever level of service is being provided at the local level is appropriate. In many cases, the local level of service (i.e., number of stations and staffing) is justified by a cost avoidance, unique local situations, or tradition.

It is easy to get caught up in taking care of the needs of employees or elected/appointed officials and fail to remember the primary goal of serving the public. Many chiefs will candidly comment that they will not attempt some changes because they would be unacceptable to groups such as the union, volunteer membership, the mayor, or the city council. With the appropriate preparation, leadership, and courage, the administration can gain acceptance for needed changes. This is not an easy task, but nobody promised that being the head of a fire and emergency services organization would be easy.

Moral Obligations of Public Roles

What special obligation does a person who accepts the position of chief or director of a fire and emergency services organization have to the public? Are there other obligations to members and the elected officials that also must be balanced in the administration's public duties?

This is an awesome responsibility because the administration will be directly responsible for the service the public receives and the safety of members. Also, in most cases, when dealing with the public, only the fire and emergency services administration will have the technical expertise to judge the adequacy of the service provided. Right or wrong, the chief has the power to bamboozle the public and politicians into believing the organization is providing great service when the organization and its members may not be operating at 100% effort and efficiency. There is no reason to think that individuals in public roles are released from traditional ethical and moral requirements, or that in public roles, the end justifies the means.

For example, in any fire and emergency services organization there will be some members who perform at a higher job performance level than their peers. It is very easy to start treating these few outstanding workers better than their peers. But remember, it takes a team of players to win the game. Each person will contribute to the best of their ability if they are properly motivated. The job of a leader is to get all employees to operate near their maximum effort, whatever that level may be.

Only in those cases where an employee's behavior is unacceptable should different treatment be administered. In these cases, the other employees will be watching, and if they see an employee getting away with substandard performance, some will lower their performance levels. Everything an administrator says or does is a signal to all employees of what to expect in relation to job performance. If the chief allows a few self-motivated overachievers to carry the bulk of the work, the chief is not treating employees impartially or fairly.

Duty to Obey

There will be many times during a career in government when a chief may be pulled in several directions when considering an administrative action. One question a chief should ask is: How strong is a duty to obey an order by a superior such as a mayor or city administrator? ". . . administrators typically work in environments where the presumption of the obligation to obey is powerful—so powerful that, in an extreme case, the administrative environment may promote a willingness on the part of officials to give up all sense of personal responsibility" (Garvey, 1996, p. 328).

Be careful in these situations. Legally, the chief may end up in the middle of a civil or criminal action. Ethically, try to do what is *right*, using some of the techniques and suggestions in this chapter as a guideline. Recent history and observations of the daily life of public officials suggest that, for many people, obedience may be a deeply ingrained behavior tendency. This is a powerful force that can override training in ethics, common sense, and moral conduct.

A person entering an authoritarian system will no longer view themselves as acting for their own purposes, but rather as an agent for executing the orders and wishes of another person. For example, the chief receives a direct order from the mayor to eliminate all overtime that is needed for minimum staffing levels. The chief obeys the order without a fight, possibly fearing loss of his or her job.

A leader of a fire and emergency services organization must find a way to stay independent by nurturing a relationship with superiors that allows for open, honest communication. Although most of this frank communication will take place in private, a superior should be approachable without the fear of any reprisals.

In the case of a fire and emergency services organization, the agency may have many rules or standard operating procedures (SOPs) that outline actions during emergency operations. An observer watching the performance at an emergency incident may criticize the organization for what they believe is an inappropriate way to handle an incident. In these cases, the problem may be the system as coordinated by the leadership, not the individuals of the organization. The chief should be approachable either personally or through a formal critique procedure to receive constructive criticism. It is very important to make this distinction when dealing with a situation that results in basic change in the organization.

Administrators will have to look out for the public because many of the groups attempting to influence the behavior of the organization will have their own agendas. Remember, these groups are made up of good people in organizations that were specifically created to look out for the well-being of their members. Groups such as unions, volunteer organizations, homebuilders, and apartment owners have all used their influence to change or attempt to change the fire and emergency services organization over the years.

Ethics of Cost-Benefit Analysis

In most cases, the presumption is that an existing or new program will not be funded if its costs outweigh its benefits. In private industry this would be the typical plan of action. The financial bottom line is one of business owners' and stockholders' most important considerations in many decisions in the capitalistic business world.

However, for environmental, safety, and health regulations, there may be many instances where a certain decision might be right even though its benefits do not outweigh its costs. There are good ethical and moral reasons to oppose efforts to put dollar values on non-marketed benefits and costs such as the value of a life.

If the chief proposed to adopt a new fire safety regulation that would require residential sprinkler systems to be installed in all new one- and two-family dwellings, the opposition would probable be able to produce a cost-benefit analysis that shows that the costs (installation of the system) outweigh the benefits (savings in fire insurance). However, the difference in these values has become almost insignificant as the cost of residential sprinkler systems continues to fall. But the opposition would probably use the highest costs it could find to help support their argument that it would be a substantial additional cost to the homebuyer.

Some of the benefits that cannot be quantified in this type of fire safety program are those of saving lives, preventing injuries, and avoiding the destruction of nonreplaceable personal items in fires.

A classic problem in selling the need for adequate fire and emergency services is the relatively infrequent need for emergency life-saving services. This makes the job of a fire and emergency services administrator very difficult, but not impossible. It is truly a challenge that must be met with knowledge, energy, courage, and planning to be successful.

Consequences

Fire Chief Fired for Lying to City Council. What would it be like to be the chief in this newspaper headline? Many times this same result occurs and there is no public realization of the circumstances. How prevalent are deceit, misleading statements, and lying in our society? The following article may give some insight into this question:

Study Finds Widespread Lying, Cheating among U.S. Teens

October 16, 2000

LOS ANGELES, California (AP)—Many U.S. high school students lie a lot, cheat a lot and many show up for class drunk, according to preliminary results of a nationwide teen character study released Monday.

Seven in 10 students surveyed admitted cheating on a test at least once in the past year, and nearly half said they had done so more than once, according to the nonprofit Joseph & Edna Josephson Institute of Ethics.

"This data reveals a hole in the moral ozone," said Michael Josephson, founder and president of the Marina Del Rey-based organization.

On the other hand, the results were not significantly worse than on the last test in 1998—the first time that has happened since the group began testing in 1992.

"The good news appears that it's peaked," Josephson said. "The bad news is that it's horribly high."

The "Report Card on the Ethics of American Youth" found that 92 percent of the 8,600 students surveyed lied to their parents in the past year. Seventy-eight percent said they had lied to a teacher, and more than one in four said they would lie to get a job.

Nearly one in six students said they had shown up for class drunk at least once in the past year. Sixty-eight percent admitted they hit someone because they were angry. Nearly half—47 percent—said they could get a gun if they wanted to.

Josephson said the results amounted to the formula for a "toxic cocktail" involving "kids who think it's OK to hit someone when they're angry, who may be drunk at school when they do it and who can also get their hands on a gun."

Josephson stopped short of assigning blame to a particular group, but he said parents, teachers and coaches need to pay special attention because they have the most significant interactions with youngsters.

"I'm not saying there aren't some out there doing their best," he said. "But if all three were doing their best, we wouldn't have this problem."

The survey, conducted this year, involved students in grades nine through 12 in both public and private schools. Participating schools handed out surveys with 57 questions that students could submit anonymously.

The results had a margin of error of plus or minus 3 percentage points.

The high school results, along with those for middle schools, will be included in a series of three final reports to be released later this year. (Associated Press, 2000)

These same high school students will be the new members of the fire and emergency services organization. They also will be the future officers and elected or appointed officials.

With the media's global reach, everyone has instant access to news about fraud, corruption, and cheating. The more immoral the act and the more famous the person, the bigger the story. A public official can no longer cover up inappropriate behavior. If the administrator has any enemies or members in his organization who do not approve of his leadership, impropriety can quickly be turned into ammunition.

For example, in reflecting on President Clinton's testimony, the judge stated "It is simply not acceptable to employ deceptions and falsehoods in an attempt to obstruct the judicial process" (Bok, 1999, xvii). The city manager may have a similar belief and feel it is not appropriate to use misleading statements to gain approval of the budget, for instance. How do you feel about the following questions?

- Should physicians lie to patients who may need an operation or are dying so as to postpone the fear and anxiety?
- Should professors and supervisors exaggerate the excellence of job applicants in order to give them a better chance at being hired? Is this fair to the employer?
- Should journalists write untrue statements or fail to tell the whole story to support their personal views on a subject that they strongly believe is right?
- Should a fire chief exaggerate the consequences (e.g., more deaths and property loss) of not funding the budget properly?
- Should a police officer use fabricated evidence to help convict a known criminal?

In the fire and emergency services business, members have many strong beliefs. In many cases they are well founded. But, they should be prepared to use truthful statements when justifying their beliefs to the public and elected officials.

For example, a department operates at an incident where there is a fire death and the chief is asked by a newspaper reporter, "What could have been done to prevent the death?" The chief could answer that having four fire fighters on each engine would have averted the death, and fail to mention several other answers that could also have prevented the death from occurring, such as smoke detectors and sprinklers. In the emergency services business there is often more than one way to prevent a tragic incident and, to tell the whole truth, the chief should mention all solutions.

Unintentional Miscommunications

Be very careful that all communications are received accurately. A person may be blamed for lying when actually something went wrong with the communications. This is especially true with verbal communications.

A person may have meant to say one thing, but actually said something with a completely different meaning. Another occurrence is when the receiver may have already assumed a certain response other than what was actually said. This is a case where a person hears what they want to hear.

"The many experiments on rumors show how information can be distorted, added to, partially lost, when passed from one person to another, until it is almost unrecognizable even though no one may have intended to deceive" (Bok, 1999, p. 16). This is a common demonstration in management training where a person tells a short story privately to one person, then has them pass the same story to the next person, until the story is passed eventually back to the originator. In many cases, the story has no resemblance to the original account.

Lying and Free Choice

When talking about taking someone's freedom away, most people think of a dictatorship or imprisonment where freedom is taken away by physical force. But deceit and lying can also result in the loss of freedom, and are common in people's personal lives and public duties. In a dictatorship, this type of communications is normally called *propaganda*.

For example, before becoming public knowledge, a well-known manufacturer of tires kept secret from the public many reports of vehicle rollover crashes involving their tires. The public's freedom to choose the safest tire for their cars was thus taken away from them by deceit. Remember, deceit does not have to be a direct communication of something that is untrue; it can be the withholding of information that others have a right to know because it involves their safety.

If asked to name this tire company, many consumers could probably do so without hesitation. This points out the long-term effect of deception: Once someone or an organization has lost the confidence of the public and members of its organization, it becomes very difficult, if not impossible, to ever gain their trust again.

There are some studies that indicate it can take up to 2 years for others to acknowledge any changes a person or an organization makes in their behavior. Once a bad reputation is created, it is an uphill battle to convince others that a change has been made for the better.

Self-Defense

"If to use force in self-defense or in defending those at risk of murder is right, why then should a lie in such cases be ruled out? . . . Surely if force is allowed, a lie should be equally, perhaps at times more, permissible" (Bok, 1997, p. 41). This supports the idea that anytime a person can justify physical force, lying can be justified.

There are some situations in which lying could be an option because physical force was justified, but a person should always consider whether telling the truth would work along with the option that physical force could be used if necessary as a final solution. However, lying to liars does pose some problems. For example, before the invasion of France during World War II, the Allies went to great efforts to deceive the Germans as to the location of the invasion. If the enemy knew the location they would have a better chance to repulse the liberation efforts. But, if the Germans were convinced that their enemies would always lie to them, then they may not have been so easily deceived. Then, when it came time for discussions on surrender or peace agreements, would the enemy believe anything said to them if they had always been lied to in the past? Even with enemies, it is often better to tell the truth.

"Alternatives still have to be weighed, moral arguments considered, the test of publicity taken into account. . . . But if, finally, the liar to whom one wishes to lie is also in a position to do one harm, then the balance may shift; not because he is a liar, but because of the threat he poses" (Bok, 1999, p. 133).

Professional Ethics

Many professions have organizations that regulate the ethical and incompetent behavior of those who are members of their organizations. In many states, a board is created to regulate the profession, and most of the board members are the same professionals. "They should expose, without hesitation, illegal or unethical conduct of fellow members of the profession. . . . Yet in practice such exposures are extremely rare" (Bok, 1999, p. 154).

The fire and emergency services profession does not have any organizations that are able to watch for unethical actions. Some organizations have the poten-

tial to make judgments on professional conduct, such as the National Fire Protection Association (NFPA), the International Association of Fire Chiefs (IAFC), the National Volunteer Fire Council (NVFC), and the International Association of Fire Fighters (IAFF). Some have a suggested code of ethics for members to guide their conduct, such as the following notable example, from the Florida Fire Chiefs' Association:

Code of Ethics
The purpose of the Florida Fire Chiefs' Association is to actively support the advancement of the fire service, which is dedicated to the protection and preservation of life and property against fire and other emergencies. Towards this endeavor, every member of FFCA shall, with due deliberation, live according to ethical principles consistent with professional conduct and shall:
- Maintain the highest standards of personal integrity; be honest and straightforward in dealings with others; and avoid conflicts of interest.
- Place the public's safety and welfare and the safety of employees above all other concerns. Be supportive of training and education, which promote safer living and occupational conduct and habits.
- Ensure that the lifesaving services offered under the members' direction be provided fairly and equitably to all without regard to other considerations.
- Be mindful of the needs of peers and subordinates and assist them freely in developing their skills, abilities, and talents to the fullest extent; offer encouragement to those trying to better themselves and the fire service.
- Foster creativity and be open to consistent innovations that may better enable the performance of our duties and responsibilities. (FFCA, 2003)

In situations where a death or serious injury has occurred, state or federal OSHA agencies are able to assess unethical conduct when an administrator, supervisor, or organization has failed to comply with a safety regulation. In cases of noncompliance, it is the organization that is generally found at fault, not the individual. So again, it is up to the fire and emergency services administrator to be ethical because it is the right thing to do, not because there will be accountability. However, there have been some cases recently where the chief eventually was dismissed after firefighter deaths were attributed to noncompliance with safety regulations or standards.

The Noble Lie

"When these three [crisis, harmlessness, and protecting a secret] expanding streams flow together and mingle with yet another—a desire to advance the public good—they form the most dangerous body of deceit of all" (Bok, 1999, p. 166). Administrators who see themselves in some way superior to the public by birth, wealth, training, or education often feel they have a right to make decisions for the public. It is a good idea to put yourself into someone else's shoes to keep from making this mistake in ethical judgment.

The administrator should attempt to picture him- or herself as old, young, a minority, male, female, a member, a customer, or an elected official. In each of these roles, how would a person feel about the choices being made for them without their knowledge? It is common for administrators to think they are the only competent person to make the decision. This leaves only their own conscience to judge whether they have made the morally right choice.

For example, an incumbent mayor is running for re-election and is convinced that his re-election is in the best interest of the city. Although it should not be a surprise, all of his staff and his political party would also support this notion strongly. Therefore, the mayor may hide an indiscretion, not be completely frank about his ideas on a controversial subject, and support a popular proposed law or spending of tax revenues that the mayor knows would not be in the best interest of the city and most of the citizens. Is this fair to the voting public?

An Example of a Tough Ethical Choice

The following is a story by Chief Edwards (retired) about his struggle with a serious ethical question that could have cost him his job:

To illustrate how ethical issues occur in the real world, I will use a personal example. During my first week as fire chief for the Prince George's County [Maryland] Fire Department I was faced with a situation that caused me concern, even though I knew what to do. A Hispanic citizen group in the county was conducting a fundraiser and reception for the county executive. The event was advertised and tickets were sold for this important political event. I learned that the event was to be held in a vacant store that had significant fire code violations. Several hundred people were scheduled to attend this event. When I first contacted the people who arranged the event they were extremely upset and wanted the event to go forward. At one point I felt as though I would have the shortest tenure of any fire chief in the history of the department.

Some members of my staff advised how to avoid the situation or get around the code for a one-time event. Others advised not to allow the event to be held. With no time to bring the building into compliance before the event, I contacted the county executive and explained that I could not allow the event to be held in that building. His response really impressed me: he thanked me for keeping him from being in a potentially embarrassing situation and that he understood the position that was necessary. The event was subsequently held in an outside tent close to where the original event was scheduled. I learned a valuable lesson in ethics and an even more important lesson about the ethics of the person I worked for. It made me feel a great deal of respect for the county executive. This example demonstrates the respect that lies within ethical decision making and its importance to your career and those who work for you. (Edwards, 2000, p. 13)

It takes courage and personal integrity to make tough decisions such as those faced by Chief Edwards.

Ethical Tests

The following is a list of a few issues for you to consider when evaluating the pros and cons of lying:

- Assume public review and discussion. (If the decision must be in secret, a discussion using this assumption will provide needed input.)
- Follow the Golden Rule—Treat others as you would want them to treat you.
- Consider the viewpoint of the person(s) deceived.
- Examine carefully the *avoiding harm* excuse.

- Investigate from top to bottom the *produces greater benefit* justification.
- Consider the *fairness* of the decision.
- Watch out for the claim of *veracity (greater truth)*.

Keep in mind this quote from Mark Twain, "When in doubt, tell the truth. It will confound your enemies and astound your friends" (Twain, 1906).

CASE STUDY #1: Automatic Sprinklers

The chief is scheduled to testify at a county or city supervisor's workshop on a proposed fire code amendment. The proposed amendment will require automatic sprinklers to be installed in all newly constructed parking garages.

An automobile fire in a parking garage several months ago prompted this proposed fire code amendment. Fire companies experienced a difficult time advancing hose lines to the area of the parking garage where the fire was and, subsequently, the fire spread to cars on each side of the original fire. After the fire was out, a reporter from a local newspaper interviewed the chief and asked why the fire companies seemed to have so much trouble putting out the fire. Without much thought, the chief pronounced that if a sprinkler system had been installed in the parking garage, the fire would have been contained with a small loss.

Some time later, the chief received a report from a battalion chief who related observations that indicated the fire company had been very slow in advancing the hose line at this fire. The battalion chief then went on to recommend that the department should institute a "back to basics" program. The BC also noted that the number of incidents of structural fires was down and the proficiency of all companies seemed to be diminishing.

In doing research to support this request, the chief discovered that the model building codes do not require sprinklers in this occupancy if there are appropriate openings on each level of the garage. The chief also found a few articles in fire publications that supported the lack of justification for sprinklers in parking garages from both real-life experience and some controlled burns.

The fire chief had been in the present job for several years and had a great reputation as a good leader and very knowledgeable. In other presentations and appearances before public hearings, the chief has found a lot of support for the department's opinions, goals and ideas.

Discussion Question

1. Describe what actions should be taken and explain any ethical considerations that would affect your actions.

CASE STUDY #2: Call Response

Your fire and emergency services agency has just experienced a 45% increase in emergency responses after implementing a new policy that requires the response of an engine company to all calls for medical emergencies.

Discussion Questions

1. Based on a cost-benefit analysis, are the citizens getting more for their tax dollars?
2. How would you defend the slow response to a structural fire when the first due engine company was busy at a medical emergency of a broken arm for a 45-year-old construction worker? At the structural fire, a 5-year-old girl died. Consider all *ethical* considerations.
3. With this new activity level, the officers of the engine companies are now reporting that they are not able to complete their in-service

fire prevention inspection and physical fitness programs. Do you think this will make it difficult for your organization to meet its goal for quality service to the public?

4. Would you consider a change in response policy that required an engine company response only on those truly life-threatening emergencies such as heart attacks, serious trauma, and trouble breathing, to reduce the call volume?

References

American Heritage Dictionary of the English Language, 4th ed. http://education.yahoo.com/reference/dictionary/(accessed July 24, 2005), 2005.

Associated Press. Study finds widespread lying, cheating among U.S. teens. October 16, 2000.

Bok, Sissela. *Lying—Moral Choice in Public and Private Life.* New York: Vintage Books, 1999.

Edwards, Steven T. *Fire Service Personnel Management.*Upper Saddle River, NJ: Brady/Prentice Hall Health, 2000.

Florida Fire Chiefs Association. *Code of Ethics.* http://www.ffca.org/ffca_administration_ethics.htm (accessed August 3, 2005), 2003.

Garvey, Gerald. *Public Administration: The Profession and the Practice.* New York: St. Martin's Press, 1996.

IAFF & IAFC. *IAFF/IAFC Two in/Two out Questions and Answers.* http://www.iaff.org/safe/pdfs/2in2out.pdf (accessed July 24, 2005), 1998.

NFPA. *Fire Department Occupational Health and Safety Standards Handbook.* Quincy, MA: NFPA, 1997.

Twain, Mark. http://www.twainquotes.com/freefind.html. (accessed July 24, 2005) March 9, 1906.

11

Public Policy Analysis

Knowledge Objectives

- Comprehend the different types of policy analysis.
- Understand the impact and influence of policy analysis on the budget and on resource allocations.
- Understand the use of statistics in policy analysis.
- Examine the necessity of consensus building and the political process used to implement policy changes.
- Analyze and research case studies and other historical experiences.
- Examine the role of the administrator as a leader and facilitator in policy analysis.

Decision Theory

How do most fire and emergency services organizations make decisions and implement public policy? This chapter aims to help answer this question and be a guide for selecting the best techniques in this process to create and implement a plan for public safety policy.

It is rare, for example, to find a fire department that has doubled its budget in one year. "Ideally, the policy analyst totes up all the negatives of a proposed course of action (the costs), all the positives (the benefits), and then repeats the procedure for all courses of action that might be taken in the circumstances" (Garvey, 1996, p. 455).

Because of the complexity of many problems in the public policy arena, the most common method for change is *incremental adjustment*. This works by making a small change and assessing the results. This tends

to be a safe way for policy makers to proceed because they do not have to risk making big changes that may have unforeseen consequences.

This can drive the typical action-oriented fire and emergency services administrator crazy. In the operational side of their public service, emergency services personnel take immediate action and complete emergency operations in minutes or hours. A long-term incremental implementation plan can take years to complete.

The most serious problem with public policy is its implicit and unquestioned assumption that policy makers control the organizational, political, and technological processes that affect implementation. As discussed in Chapter 3, leading change is not a simple process. The notion that policy makers exercise, or ought to exercise, some kind of direct and determinant control over policy implementation might be called an *erroneous belief* of conventional public administration and policy analysis. However, once a new policy is identified, techniques discussed in Chapter 3 will help implement change.

Strategic Planning

Strategic planning is normally done at the department level. This type of planning identifies the department's vision, mission, and strategies. The U.S. Fire Administration (USFA) document *America Burning* has some good guidelines for the vision and mission of a modern fire and emergency services organization. Generally, an appointed task force of fire and emergency services officials and senior members use this method of planning. In some cases, an official master plan will be produced with input from the public, other agencies, and appointed/elected officials of the municipality. Strategic planning can help identify where the organization is at the present time and where a new vision can take the organization in the future.

Next, to implement the strategic plan, tactical plans should be prepared and monitored on a monthly basis. For example, in a situation where the department has just received additional funding for a new station and part of the justification was a reduction in response time to a specific area of the jurisdiction, there should be a monthly check of response times to emergency incidents. A formal report should be prepared to brief the executive and elected officials. At the appropriate time, a public summary of the new program should be prepared to inform citizens and the press of the impact of the new spending of tax dollars.

In addition, there are operational planning items that are done on a day-to-day basis. For example, it is common to have minimum staffing levels that must be maintained daily. There may also be guidelines for the number of people who may be on annual leave for any particular day. If staffing levels fall below the minimum level, personnel must be called back on overtime. It is very important to document the exact reasons for unexpected expenses like overtime. At the tactical planning level, this information can be used to suggest changes in fire and emergency services departmental policies. For example, if the timesheets show an increase in overtime for personnel using sick leave, an evaluation of leave policy and decreasing the number of personnel authorized to be off on annual leave may be appropriate.

One good example of the thought process is outlined in the following model:

Rational-Comprehensive Decision-Making Model:
- Process of making decisions based on logical reasoning
- Accepts significant change to reach the final solution to a problem
- Favors the comprehensive solution to completely resolve the issue as opposed to temporary "quick fix" measures
- Uses a dispassionate and orderly thought process to:
  1. Identity problem (not a symptom).
  2. Rank goals.
  3. Identity all possible alternatives that solve the problem.
  4. Perform cost-benefit analysis.
  5. Perform comprehensive analysis of various ways of solving the problem.
  6. Select alternatives that accomplish goals. (Lindblom, 1993, p. 147)

Linking Policy Analysis to Budgeting

The budget and planning processes are a continuous cycle. The plan should dictate the needs that are placed in the budget. When the planning process is completed, the administration should determine the financial resources needed to achieve the goals. Creative strategies for funding other than tax revenues may be available and should be pursued. This is always a good way to gain respect from elected officials.

If the numbers in the budget become the focus of the analysis, administrators will find they are simply adding an incremental percentage to each line item for the proposed budget each year. If a strategic plan is followed, the chief may decide to move funds from one part of the budget to another to fund a new goal, priority, or objective, or there may be a justifiable need for a significant increase in the budget.

For medium to large fire and emergency services organizations, budget preparations should start at the lowest feasible level of management. For example, the chief may direct the fire prevention division to prepare its own budget based on an incremental increase over last year. In addition, division chiefs may be asked to submit any justifiable enhancements as separate budget items. Finally, division chiefs should prepare a budget with a percentage decrease, including identification of programs that could be eliminated. This will give the administration many options to fine-tune the budget. The chief may be able to eliminate funding for some items while at the same time adding funds for new programs.

At this stage, the division administrators may be asked to prioritize between adding back the proposed percentage cuts and funding the enhancements that have been requested. As the next step, combine all the proposed budgets from all divisions along with their proposed decreases and enhancements, and have the entire group prioritize any decreases and the enhancements.

Once the fire and emergency services staff have completed this prioritization, it is now up to the fire and emergency services administrator to balance funding by choosing decreases that are justified, adding funding for those enhancements that warrant support, and requesting additional funds for other enhancements that the administration believes are justified. It is now time to prepare a presentation for the elected and appointed municipal officials' review. Remember that these officials must be the arbitrator between the competing agencies' requests for public funds, so the chief should prepare the best justification and present it in a style that is easy to understand.

Whenever possible, use common sense arguments and analogies. For example, if the department was requesting funding for a new fire station and fire fighters to staff the station to reduce response time in one section of the city, the administration would want to determine the response time in the other areas (**FIGURE 11-1**). Then make the common sense argument that all residents in the city deserve equal fire and rescue protection (similar response times); therefore, this proposal will bring xx% (or an actual number) more of the city's residents into the adequate response time area. Also, a picture is worth a thousand words and should always be used.

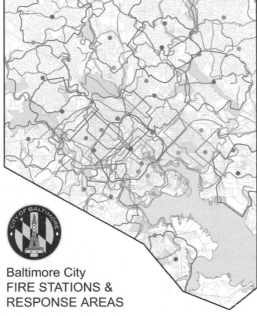

Baltimore City
FIRE STATIONS &
RESPONSE AREAS

FIGURE 11-1 Four-Minute Response Contours

Budget Process Benefits

Including managers of the major divisions in the organization in the preparation, justification, and prioritization of the budget will help reduce internal conflicts between the different areas of the organization. This will help these managers see the big picture and be able to support the unified budget. Remember that there may be as much competition for the budget dollars internally as there is externally.

Using groups of people to study and make recommendations about the organization's goals and objectives can be a very effective administrative tactic. First, the administration will seldom be blindsided by the lack of appropriate information and expert advice when analyzing a management or functional problem. Also, after the decision has been agreed to by consensus, expect to receive the support of all the managers.

What Are Outcomes?

An *outcome* is usually a measured benefit to stakeholders created as a direct result of dollars spent by the organization. Many of the measurements that are kept by the traditional fire and emergency services organization are measures only of workload, not of outcomes. For example, fire and emergency services organizations keep accurate statistics on the number of emergency incidents to which they respond, but that does not show how many lives and how much property was saved, which would be an outcome.

The U.S. Fire Academy's Degrees at a Distance course book, *Advanced Fire Administration: Course Guide* (1998) contains the following observations: "How do we measure lives saved, fires prevented, and property damage minimized? They are all basically immeasurable. . . . How can prevented incidents be recorded in order to be counted in the statistics? Not having fully measurable results is a problem with most service-oriented agencies" (USFA, 1998, p. 4-2).

Assessing the number of lives and properties saved is very difficult, if not impossible. For example, if the department responds to a small trashcan fire in a high-rise building and extinguishes the fire quickly, could the department claim to have saved the entire dollar value of the high-rise building? Some organizations do. However, the department would have a better claim of saving the entire building if a complete floor was on fire and it was successful at extinguishing the fire before it extended to any other floors.

It is also rare to be able to document saving a life. If fire fighters were to search, find, and remove an unconscious person from a building on fire, they would be able to claim an unquestioned save. In the fire prevention arena, there are more opportunities to claim lives saved from built-in fire protection equipment such as when a smoke detector activates and the occupants escape without injury. Many saves, however, are never reported to the fire department. A survey of customers may be the only method to assess the real impact of many built-in fire protection items.

Outcomes for a fire and emergency services organization must be measured using other items that are easily and verifiably measured. For example, there should not be disagreement with the common sense assumption that the sooner the emergency responders arrive, the better chance that property and lives can be saved. If it can be demonstrated that response times can be reduced by either changing existing procedures (e.g., streamlining the call-taking process) or adding new resources (e.g., increasing the budget for new staff or new facilities), then the argument can be made that improved service to the public has been accomplished. Even in volunteer organizations, implementing a system to have members on duty in the station will substantially reduce total response time and would have an excellent outcome.

The following two items can be used to measure outcomes only indirectly, but they are entirely valid.

Response Times

The number of fire stations serving an area has the greatest impact on response time for a fire and emergency services department. If a community were to construct one fire station for each block, the best response time that money could buy would be realized. But this is unrealistic. Some jurisdictions use their existing fire station spacing as the local benchmark. Paid fire fighters are used to staff units that respond immediately, keeping response times to a minimum. Some volunteer-staffed companies also may respond immediately if an appropriate number of members are in the station. A few volunteer organizations require members to *stand by* in the station to provide an immediate response. The recently adopted National Fire Protection Association (NFPA) Standard 1710 has now set a benchmark for response times. After dispatch, the first due company has 1 minute to start its response and 4 minutes for travel time to the emergency incident (for a total of 5 minutes).

The experience of adjoining and nearby jurisdictions and national surveys also can be reviewed and considered in the response time analysis. For example, a 1998 national survey by the Phoenix Fire Department contains the following results for average response times:

Population	First Unit to Arrive (minutes:seconds)
0–99,999	3:41
100,000–499,999	3:56
500,000–999,999	4:02
1,000,000+	3:59
Total average	3:50

Note: "These results reflect the experience of 335 fire departments protecting 82,339,257 people . . ." (Phoenix Fire Department, 1998, p. 7). These are averages that include turnout time and cannot be directly compared to the NFPA response times of a total of 5 minutes for 90% of the calls, which is a maximum standard. It would not be surprising if most of these departments could meet the NFPA standard; however, some may have to relocate some stations for more uniform distribution.

Staffing Levels of First Unit to Arrive

Several studies have been conducted that relate efficiency, measured by the time required to complete fireground operations, to staffing levels. Also, the minimum number of personnel is addressed in the Occupational Safety and Health Administration (OSHA) Respiratory Protection Standard 29 CFR 1910.134 (two in/two out) and NFPA Standard 1500. This minimum number is four personnel. Anything less than this number will cause delays in attacking the fire and/or unsafe actions by fire fighters. In addition, NFPA 1710 is based on the recommended staffing of each company with a minimum of four fire fighters.

Because some outcomes are relative items, they can often be evaluated by looking at efficiency, responsiveness, or equity. For example, a new computer-aided dispatch system may have been purchased that, along with new policies and upgraded training for dispatchers, will reduce the dispatch time by 50%. This measured outcome is an indication of greater efficiency, greater responsiveness, and equity because it serves equally everyone who calls for emergency assistance. This is a great outcome as a result of the improvements in the dispatch system.

In general, again using emergency response to a structural fire or medical call as an example, many improvements cannot be measured directly (saved three people last week); however, relative improvement can be measured. A better outcome is the result of arriving sooner, starting emergency operations sooner, and using professionally trained and equipped responders.

Increased Productivity

"The fire service in the United States knows that it is doing a better job and most of its indicators, such as fire deaths, fire injuries, number of structural fires, and others, are all down from previous years" (Edwards, 2000, p. 246). This statement is very true (except regarding fire-fighter deaths) when looking at national statistics from 1992 to 2001 that reflect a 10-year trend. For example, the following decreases have been noted: fire incidents per million population (PMP) are down 23.9%, civilian deaths PMP decreased 29.6%, fire-fighter injuries are down 17%, fire-fighter deaths PMP are *up* 30%, and direct property damage adjusted for inflation is down 6%. Fire and emergency services professionals must remind themselves that although great progress has been made, there clearly is more work to do. (USFA, 2004, p. 1-11)

The following are some trends that may be affecting these national fire statistics:

- Smoke alarms have been installed in 88% of U.S. homes. This has had a great effect on lowering fire deaths. It has been increasingly noted, however, that approximately 60% of fire deaths occur in homes where the smoke alarm is installed improperly, is dysfunctional, or is missing the battery. This statistic also points to the possibility that many homes protected with working smoke alarms may be having fires, but the fires are detected early and extinguished by the occupants, and the fire department is never called.
- The number of cigarette smokers in the country's population has stayed relatively level over the past 10 years. Going back to 1965, 42.4% of the adult population smoked, whereas in 1995, the smoking population was 25.5%. The actual number of smokers has decreased by only 6% since 1965, but the percentage of the population has decreased more because the population has increased. The greatest percentage of fatal fires are started by misuse of smoking materials, so this may have had a big impact on fire deaths over the past 35 years.

- Mandatory installation of residential sprinklers continues to raise strong opposition and receive little support from homebuilders and elected officials. As a result, fire sprinklers protect few single-family homes. However, at its annual meeting in 2005, the NFPA membership and committees supported NFPA 101, *Life Safety Code*, which adopted a requirement for residential sprinklers to new one- and two-family dwellings.
- Although hard data are not readily available, inadequate staffing levels in fire companies have been pointed out as the cause of increased property and life loss along with fire-fighter injuries and deaths.
- The addition of first responder medical service by fire departments is very common. This increases productivity because the same people are doing fire and EMS. In most cases, citizens are receiving medical treatment a lot faster because the response times of fire companies are shorter than for the typical EMS unit. However, this can increase the response times for fire calls because the closest fire company may be working at an EMS incident when a fire is reported in its first due area.

As a general guideline, if the fire company responds to over 2,500 calls per year, look very closely at the need for a second unit, either fire or EMS. Another useful measure is to do a time motion analysis. Actual response data should be used, including the time out of service spent either responding to or working at emergency incidents. If it totals over 10% of the available time during the busiest 8-hour shift, it may be time to plan for a second unit. In policy analysis, be very careful when considering productivity increases that reduce other measured outcomes, such as response to fire emergencies.

Statistics

Statistics can be very useful for creating and justifying policy analysis and recommendations. In addition, statistics can be used to help manage day-to-day operations. However, as with the term *outcomes*, statistics can be misused and misunderstood.

FIGURE 11-2 shows the plotting of data from many structural fires. Looking from left to right, it is evident that the average dollar loss per incident goes up as the number of fire fighters who responded to the

Average Property
Loss per Fire

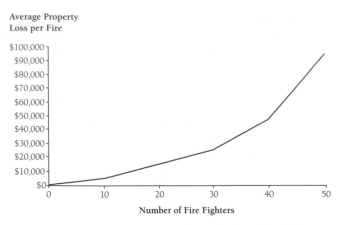

FIGURE 11-2 Average Property Loss per Fire vs. Number of Fire Fighters Responding

incident increases. No surprises here. These data are an average taken from the experience of numerous fire calls. For example, a call with a million-dollar loss may require 100 fire fighters on the scene. Fire calls with minimal property loss require only a few fire fighters to extinguish the fire.

Let's say that an enlightened city manager is using this data and graph to determine how to reduce fire loss. The city manager notices that reduced property loss is directly associated with fewer numbers of fire fighters that respond. Therefore, the obvious answer is to send very few fire fighters to fires and there will always be small losses. The city manager can prove this conclusion by looking at the graph (statistics) and noticing that when no fire fighters responded, there was no loss. This is what the statistics clearly show.

This is a good example of how the data can be correct, but the analyst neglects to use common sense tests of the results. Another common misused statistic uses the analogy that because most fires are extinguished by using small amounts of water, the department only needs one or two fire fighters to put out fires. However, when looking at these statistics more closely, the analysis will conclude that a small percentage of fires cause the greatest number of deaths and the highest amount of property damage. Therefore, from a quality service perspective, the department should be prepared to handle these challenging fires that cause the greatest amount of property loss and deaths. A professional fire and emergency services department should be able to give a 100% effort 100% of the time. This should be the goal to assure quality service to the public.

What does it mean to the public if a business or agency gives less than 100%? Paul F. Wilson, Larry D. Dell, and Gaylord F. Anderson (1993) in *Root Cause Analysis: A Tool for Total Quality Management*, reported

the following results if goals are lowered only .1% to 99.9%:

- Two unsafe landings at O'Hare Airport every day
- 16,000 lost pieces of mail per hour
- 20,000 incorrect drug prescriptions each year
- 500 incorrect surgical operations performed each week
- 50 newborn babies dropped at birth by doctors each day
- 22,000 checks deducted from the wrong account each hour

Another more recent use of statistics concludes that because the greatest majority of emergency responses are for medical calls, the fire and emergency services organization should concentrate its efforts on these responses. Again, this analysis tries to discount the importance of being able to extinguish a truly challenging fire before it can cause extensive damage and deaths. In reality, the fire and emergency services organization should be able to give 100% professional effort in extinguishing structural fires and also be able to give the same effort in providing emergency medical services. In other words, the fire and emergency services organization should be professionally competent to provide the best service for both fire suppression and emergency medical services. Neither one should detract from the other. If these goals cannot be met, the administration needs to ask for additional resources or limit service to only one emergency type.

Statistics from studies or surveys are very common today. Reporters of the statistics can put their own twist on the results. If administrators and/or their staffs do not have a good understanding of statistics, they can be unintentionally misled by some of these reports.

Consider the story of a race between two horses, one American and one Russian, which took place in the Soviet Union. The American horse won the race, and as the story goes, it was reported in the Moscow newspaper as follows: "The Russian horse came in second place and the USA horse finished next to last." The statement is accurate, but very misleading. Although this is a simple example, the same kind of reporting of findings from a survey or study is always possible.

A complete understanding of statistics and all of their idiosyncrasies cannot be achieved in the amount of space available in this book. Most computer spreadsheets can do simple statistics from the results of a survey. Be careful in relying on these results, however, because they are very simple tests.

There are several items to look for in a reliable survey. The first is that the survey was truly random. Most surveys sample a small number of people and then generalize the results for a large population. One method of randomly selecting participants is to use a random number generator (on a computer) to select telephone numbers from all possible phone numbers in the area being surveyed.

A sufficient number of respondents are needed to assure that the *standard of error* is small. Not everyone is being surveyed in the population for their answer(s), so the results are a prediction of the entire population based on a small sample. The more people surveyed, the better chance you have of getting an accurate prediction of the real outcome. For simplicity, the survey should include at least 50 people and preferably 250 people.

It is always a good idea to review the actual survey and any assumptions used in the study. For example, a study of fire fighters and the incidence of heart/lung illnesses concluded that fire fighters had a much higher incidence of heart/lung problems than the general public. This study was used to justify most of the heart/lung presumption retirement and health benefits that are common throughout the country.

That same study was redone except it was *normalized* for cigarette smokers. Normalized means that the study was changed so that only like situations are studied. Some studies do this by using a technique called multiple linear regression. In other words, if more than one variable could cause the outcome, in this case heart/lung illnesses, all potential variables can be studied statistically at the same time via a computer program. Independent variables are rated separately on the possibility that they caused the event, and then the chance that several variables may be working together to cause the event is also calculated.

After redoing the study by separating the fire fighters into smokers and nonsmokers, the new study found that nonsmoking fire fighters had a statistical chance of heart/lung disease slightly greater than the general public. However, the smoking fire fighters had an incidence rate many times that of the general smoking public. This suggests a synergistic outcome by combining cigarette smoking and firefighting.

Check out the source of the statistics. This does not guarantee that a structural statistical mistake was not made in the study, but it can give a strong indicator of whether the statistics were correctly reported. Be suspicious of statistics from organizations or people that have something to gain from the study. Although some of the studies are accurate, many others are structured to look at the preferred outcome and may be narrow in their focus. A 1997 survey by *USA Today* reported that 24% adults said lying is sometimes justified. These are the same adults who may be creating and conducting a statistical study.

Remember that each valid statistical conclusion from a random sampling study will have a confidence level reported, as in the following hypothetical example: "There is a 95% certainty that the average age of fire fighters in this country is between 36 and 39 years old." Or, a statistic may report that there is a plus or minus variance of 4%. Nothing is completely certain unless a study asks all fire fighters their age.

The old saying of not comparing apples to oranges is easy to understand for fruit, but hard to recognize in a statistical study. Think about the study and its conclusion using common sense. Wendy McElroy (2004) cautions about the use of statistics in public policy analysis by stating, "Our society rewards those who construct problems. They receive financing and media attention, write books and become 'experts.' Statistics are tools and those who wield them should be neither glamorized nor ignored. But they should be required to answer basic questions before being included in the rare category: purveyor of truth" (McElroy, 2004).

Consensus Building

To be successful at maintaining the department's budget and adding enhancements when needed, it is very important to understand the approval process. From the politician's point of view, spending tax revenues is an opportunity to support their personal projects and influence voters to support their election efforts. Therefore, if politicians support better education and/or can gain votes for their next election by supporting an increase in spending for education, that is what they will do.

There is always a lot of gray area when allocating tax revenues for public goods. It is the chief's job to acquire a fair share of the total budget. Although it is rare that one issue dealing with fire or emergency services would make a difference in an election, it is always seen as a positive step in gaining more votes to support popular fire and emergency services improvements.

In general, if the administration can make arguments that are backed up by common sense and solid research, there is a good chance of receiving approval for the budget request. If a national consensus standard can be used, such as those of the NFPA, to support the budget request, it is easier for politicians to support the request when they can justify their votes. Remember, with so many people requesting additional funding for their projects, the chief needs to give the municipality's administration and elected officials some solid justifications for allocating scarce resources to the fire and emergency services department.

In the fire service, there has been some controversy over the use of national consensus standards, especially the NFPA's. On the one hand, local fire and emergency services officials say that they want the freedom to provide the level of public protection that is determined necessary in the local community. However, generally, the local level of fire and emergency services protection has been created without the input of the local customer. In addition, in most cases there was no planning effort by the community or its citizens to determine the level of protection. Whatever resources and types of organizations exist at the local level are the result of a very complex and independent evolutionary process. In some cases, the type of local municipal government has an effect on the local fire and emergency services.

Fire and emergency services are in the process of entering a new era. In the emergency medical service field, responders are now certified to national standards for Emergency Medical Technician and Paramedic. In the fire service arena, national standards for training and safety have been available for a relatively short period of time, but are having an increasing influence on emergency services. These national standards can be an opportunity to justify the existing budget and/or gain approval for a budget increase.

For example, the chief may propose a new program to certify all officers to NFPA 1021, *Standard for Fire Officer Professional Qualifications*. This may require a training program, incentive pay, and overtime pay to cover classroom attendance. The selling point for this proposal is that, when finished, the city or county will have a more professional fire department. This is the kind of proposal that the municipal administration and elected officials can clearly understand.

Remember that special interest groups (for example, fire fighters' unions and volunteer fire associations) can have a big influence over elected officials. Approach these organizations and their leaders at the earliest opportunity to gain their support.

The Process for Consensus Building

In a fire and emergency services organization, consensus building must start with the municipal administrator or elected officials. This process should not be attempted if there is a feeling that these officials will perceive it not as a consensus building technique, but as a conspiracy to usurp their power. In other words, before starting, the chief should get their full advice and consent for using this technique.

In consensus building analysis, influential private citizens and any administrators who give trusted advice to the chief administrative officer, such as a budget director or public safety director, should be included. Don't forget that the staffs in other agencies have a lot of influence over their administrators. Because this can vary greatly from organization to organization, these are only generalities. The task here is to build support among these influential people so that the proposal will be approved and funded.

This technique starts off with preparing a list of all those people and groups who could influence the final decision (**TABLE 11-1**). Then, in a second column, for each individual or group, rate their potential impact on the final decision. For example, in a mayor-council form of government, the mayor and the most influential private citizen members should be rated as having a very strong potential influence. The union president may also be listed, but because of some prior

Table 11-1 Consensus Building Example		
Person or Group	Potential Influence (high, medium, or low)	For, Against, or Neutral
Chief administrative officer		
Mayor		
Power elite member (*your choice*)		
Fire fighters' union		
Local newspaper		
Police chief		
Budget director		
The public		
City attorney		

misunderstandings and conflicts with the chief administrative officer, the union official's influence may be rated as very weak.

In the next column for each person or group, the person's probable support of the proposal should be assessed. Again, for example, if the proposal were to hire an additional 100 fire fighters, the union president would be rated as strongly supporting.

Once this list is finished, the chief would seek out those people who are very influential and will more than likely support the request. In private one-on-one discussions, these likely supporters would be briefed on the proposed request, asked for their input, and then their support would be requested. This technique would also be used on individuals rated as being neutral on the issue. The object would be to bring these people to a point where they would mildly support or not oppose the proposals.

Finally, approach those persons who are rated as probably opposed to the proposal. The object here is to present the best case possible for support of the proposal, and hope that these people will become neutral on the issue. This also keeps these individuals from feeling like they have been backstabbed by their opponents without a fair chance to influence the decision. Avoid making permanent enemies; it can come back to haunt an administrator in the future. In summary, each person or organization is personally contacted by the time you finish.

This specific detailed process should be necessary only for those potentially controversial and expensive enhancement items in the proposed budget. Keep in mind that for every new tax dollar the department receives in its budget, an equal amount must be taken from another agency's budget or generated from increases in taxes. Neither option is a pleasant task for elected officials.

Public Policy Presentations

The following are some suggestions for handling public policy presentations as professionally and successfully as possible:

- *Consider the audience:* What is their real knowledge of fire and emergency services operations? It may come as a surprise, but most appointed and elected officials have no background in fire and emergency services. Will these officials understand any jargon or acronyms that are typically used? Before scheduling a policy or budget hearing appearance, it would be a good idea to arrange a tour and orientation for the officials at one of the department's stations. Use an educational program and adapt it to the adult audience. Have them try on the protective clothing and the self-contained breathing apparatus (SCBA). Explain the environment that the fire and emergency services personnel will be operating in when performing emergency operations. Use worst-case examples (although plausible) to illustrate those incidents when the department will be stressed to its maximum effort. Avoid examples that are totally unbelievable, such as suggesting the department must be prepared for the mass casualty incident when a 747 airplane crashes into the high school football stadium during a homecoming game. If the chief is unsure of elected and appointed officials' understanding of the budget request and its impact on service efficiency, use a presentation to explain the details along with emphasizing the improvements. Never assume anything. If in doubt, ask questions.
- *Practice:* There are many levels to the practice phase. At the first level of preparation, the chief should be familiar with the entire budget and justifications. One technique that has merit is to ask the staff to review the budget and write down questions that the appointed and elected officials may ask. For questions that require an in-depth knowledge to answer, an appropriate staff member should prepare a briefing document for review. This person should also be present at the budget hearing to fill in details or correct any misunderstandings. If the chief lacks presentation experience, he or she may want to practice the presentation using a friend or staff member, or videotape the simulated presentation and review the recordings. There are two different types of situations to practice—the prepared presentation and the response to the questions at the public hearing. Practice is the only way to be confident, relaxed, and persuasive and to overcome fear. As is true with any presentation, be very knowledgeable about your subject.
- *Be positive and cheerful:* Remember that politicians love to take credit for new and improved programs for the public (voters). Show them, in simple terms, the positive

aspects of the proposals. From practice, the chief should be able to maintain a positive, confident appearance that reflects professional ability and credibility. Stay away from arguments such as "If the council doesn't approve these requests, many children will die in fires this year." State the justification in positive terms. For example, point out that the department will be able to provide better service with the approval of the enhanced budget requests. Be cheerful. If unsuccessful this year, there is always next year.

- *Avoid reading and maintain eye contact:* Reading is distracting and prevents the speaker from maintaining eye contact. The presentation will have more impact and be more believable if the speaker can give the presentation extemporaneously from notes, rather than reading through pages and pages of material. Maintain as much eye contact with the elected officials as possible, because many people relate eye contact to honesty and sincerity. Scan the officials without staring. A glance at their eyes and facial expressions may give an indication of their reaction to the presentation, and the presentation can then be adjusted accordingly.
- *Use down to earth ideas:* Whenever possible, use an analogy or a real life story, perhaps including a humorous anecdote, to help explain a theoretical or abstract proposal.
- *Use visual aids:* A picture is worth a thousand words. Include visual aids such as handouts, charts, slides, and videos. Use them in a way that will not distract from the main presentation, however. If using films or videos, be careful not to use too much of the allocated time. If necessary, review the film or video and limit it to no more than 5 to 10 minutes. Selective use of video aids can increase the understanding and attention of the elected officials.
- *Involve the audience:* The appropriate techniques for this phase may vary, so having a good understanding of procedural rules, past practices, and respect that will be expected by the elected officials is very helpful. At a public hearing, elected officials will be performing for the public and press who may be present. A direct confrontation or disrespect can cause a loss of support both at this hearing and at future hearings. Research the pro-

fessional and personal backgrounds of each elected official. A thorough understanding of the elected officials and their past experiences increases the chances of involving them positively in the discussion of the budget. In addition, the chief should have had personal one-on-one meetings to get to know all the officials. At the end of the presentation, ask for questions. This is when a chief will really have to think on their feet. Remember that the elected officials may have prepared questions for the hearing. Answer truthfully and as simply as possible. Keep to the facts and professional opinion. Do not guess! If the answer does not come to mind, say so and offer to locate the answer after the hearing. One last point: A speaker should not, under any circumstances, lose his or her temper, regardless of how foolish or malicious the question or comment from the elected official seems.

- *Don't preach:* Convince the audience rather than beating them over the head with righteous logical reasoning. Sure, the chief is the most knowledgeable person in the community, and elected officials should understand and believe the chief's advice without question. But, that is not the nature of the political process. Consider the appointed administrators and the elected officials as being the *devil's advocates*, whose job it is to question everything the chief does, no matter how accurate and trustworthy the justifications are for the budget. Again, do not take any of this as a personal attack; it is normal business.

Using Case Studies

Case studies are a good approach to studying, reviewing, and analyzing administrative and policy problems. What is especially important about this method is that administrators can learn from others' experiences. It is always better to learn from someone else's mistakes or successes. This is the impetus for many case studies. And concerning our own mistakes, everyone makes a few; the trick is to not make the same mistake twice.

"It's essential for our public servants to feel comfortable and intellectually confident when deciphering a table of figures or interpreting a simple algebraic description of a problem" (Garvey, 1996, p. 6). Statistics and other mathematical descriptions of some problems in the public sector can be very helpful for understanding the problem, analyzing possible solutions, and finalizing the plan to solve the problem. For example, after reviewing data from fire deaths for the past few years in the jurisdiction, the analysis shows that the majority of the deaths occurred at night when victims were asleep. This has also been demonstrated through the review of national statistics. This type of mathematical description of a real problem can be used to point out to legislators the need for a new ordinance requiring smoke detectors in existing dwellings to solve this problem.

The chief must also keep in mind what is called *human engineering*. When dealing with human problems, administrators must consider subtle factors such as how people feel about an issue or how they will behave. In many cases, without being able to prove these factors scientifically, as in a controlled statistical study, administrators must treat these human engineering dilemmas as facts. In the case of smoke detectors, a human engineering problem is that many of the smoke detectors that have been installed were found to be nonfunctional because of missing or dead batteries. In addition, recent research indicates that some children do not even wake up when a smoke alarm activates. This is very troublesome to many fire protection professionals. (Some recent studies suggest that younger children may react better if, instead of a typical alarm sound, the voice of one of the parents is recorded and activated when the smoke alarm starts.)

"Most important of all: Trust your own experience and instincts! All this leads us to a curious conclusion: In a sense, you probably already have most of the answers (although it may take some effort to dig them out of your own store of basic instincts and ethical intuitions)" (Garvey, 1996, p. 25). In other words, what the reader should learn from this text is to be able to ask the right question and to find the best answers. But more importantly, to convince others to implement the changes or fund the improvements, people must first believe in themselves.

Formal Policy Analysis

There have been many attempts to develop ways to estimate the outcomes of official actions. As pointed out previously, it is very difficult, if not impossible, to measure direct outcomes from the actions of a public safety organization.

"As far as many policy makers are concerned, advice that is merely streetwise or commonsensical won't cut it anymore" (Garvey, 1996, p. 357). More and more, senior public administrators and politicians are demanding analytical, quantitative justifications to support budget requests and policy decisions. In the fire department, this trend can be observed in the traditional fire chief's opinions not carrying the same influence as they once did in the good old days.

One major task as a leader of a fire and emergency services organization is to gain increases in the budget for a new program or goal. Formal policy analysis can provide the justification needed when faced with scarce resources. Could you convince the elected officials to either transfer funds from other departments (more fire fighters, less teachers) or to raise tax revenues?

> [M]ost governments begin with control and its line-item form [of budgeting]. . . . [G]overnments face the problems of preventing financial improprieties; and of limiting agency spending to authorized levels. For this reason, government budgeting inevitably begins with indispensable efforts to promote "accountability" by preventing public funds from being stolen, used for unauthorized purposes, or spent at uncontrolled rates. . . . The management orientation, which marks the second stage of budgeting, is reflected in the movement for performance budgeting and the concomitant redesign of expenditure accounts, the development of work and cost measures, and adjustments in the roles of central budgeters and in their relationships with agencies. . . . [Program budgeting] represents the third stage of U.S. budgeting, the planning orientation. In this stage, the determination of public [policy] objectives and programs becomes the key budget function. (Schick, 1971, p. 9)

In many local governments, this final stage, program budgeting, is not a formal part of the budget process (which is normally a line-item budget), but a separate policy analysis. A distinct advantage in the budget battle can be expected if these techniques to support the budget requests are used. If justifications for the budget requests are supported by irrefutable quantitative outcomes, it will give the fire and emergency services administration an advantage that may bring increased funding to the agency. However, with politics there are no guarantees. In any case, the administration will have made its best effort to get the resources needed to accomplish the organization's goals.

This third stage (program budgeting) will be discussed at length in this chapter. This is where the modern public administrator can increase the quality of existing programs and gain approval for new or expanded programs. One of the major tests of the success of a leader comes at the budget table. Can the administration get its fair share of the tax revenues?

Cost-Benefit Analysis

"The techniques of formal policy analysis, with cost-benefit analysis at their center, have today become standard intellectual equipment, particularly for public administrators serving in the roles of policy analysts and advisers" (Garvey, 1996, p. 379). In many cases, cost-benefit analysis is what the typical city or county administrator relies on to provide guidance in allocating public funds and developing public policy.

In most cases it is very difficult to measure benefits or direct outcomes from fire and emergency services programs, so we will discuss a few benefit evaluation techniques that may be helpful. For example, how does one assign a cash value to a person's life or quality of life?

"The most difficult cost-benefit analyses of all are those that involve value-of-life determinations. . . . How can a public health or public safety official put a dollar value on the lives that might be saved by a new immunization program, a changed traffic pattern, or an investment in upgraded emergency medical services?" (Garvey, 1996, p. 381). If an administrator were to survey the public about the value of fire and emergency services, there would probably be a great variance in the valuations from person to person, situation to situation, and group to group.

The following discussion will seem familiar as you have already read it in Chapter 4. These examples have a dual purpose—financial and planning—therefore they are repeated in this chapter. Remember the family enjoying a nice quiet evening in front of their TV with no fire or smoke in their living area. This family would probably express a very low value in having the public fire and emergency services organization ready to respond and rescue them in a fire emergency. Take this same family and place them in a 13th floor apartment with a fire blocking their exit to the stairway, and the result would be a new higher value for the public fire and emergency services organization.

Again as discussed in Chapter 4, what makes cost-benefit analysis even more difficult for fire and emergency services organizations is the realization that the outcome—saving lives and property—is in most cases the result of a complex set of circumstances. The following is a hypothetical set of circumstances that led to a fire death in a house fire:

1. Smoke alarms were not required by the local fire prevention regulations to be installed in existing dwellings.
2. Home fire inspections along with fire safety education were not provided by the local fire and emergency services organization.
3. The occupants of the dwelling did not see any value in installing smoke alarms.
4. The occupants of the dwelling did not recognize the hazard of careless smoking in the home.
5. The first arriving fire and emergency services unit was understaffed and/or personnel were not adequately trained.
6. When the victims were eventually rescued and brought to the front yard, the fire and emergency services organization was not equipped or trained to provide emergency medical services.

It could be argued that if any of these six points were different, lives could have been saved. This makes it almost impossible to construct a valid cost-benefit analysis. Although some of these items are clearly more expensive than others to correct, it would be good to implement programs to correct all the deficiencies indicated by each item. Anything less than 100% effort in all areas will leave open opportunities for preventable life and property loss in this example.

Program Analysis

One method that can be used in place of cost-benefit analysis is program analysis. This is a systematic process to examine alternative means, methods, or policies to accomplish the goals and objectives of the fire and emergency services organization. This type of analysis accepts the use and reporting of intuition and judgment as well as statistical studies and sensitivity analysis. One caution: This method should be used only for those demonstrated problems that will have a big impact on the services the department delivers.

Program analysis works best when the chief appoints a task force representing all members of the department. This technique is used to make sure that all possible solutions are discussed and considered. There may be cases when citizens, special interest groups, and elected officials would be requested to participate. This can be used as part of a strategy to gain support for a new program.

The administrator of the fire and emergency services department will be the leader and define the problem to study. For example, a report on response times in the city notes that one geographic area has substantially longer response times to emergency calls than the averages in other parts of the city. The task group is called together, the problem is outlined, and the group is asked to complete an analysis and recommend several solutions. This group is given the following guidelines:

- *Brainstorm:* Allow all ideas to be listed and considered. Identifying the problem can be the most critical point in this process. Many times a symptom is confused with the real cause. All solutions should be analyzed. Have the group consider all traditional and inventive ways to measure outcomes.

- *Evaluate:* This is the analysis phase. Remember to test each solution for political sensitivity, human engineering problems, and common sense. Review any public preference surveys that have been conducted (see Chapter 6). Surveys can be a strong justification to support the plan.

  The following are examples of the results of evaluation: After studying existing response times in a jurisdiction, the need for a new fire station has been identified so that all citizens have access to the same level of service, as measured by response time. Or, the need for a new fire station in this identified area is not recommended because the analysis has concluded that longer response times in rural areas are sufficient. In some cases, several alternatives may be listed that achieve the same results. All solutions should be justified by valid measurable outcomes (e.g., response times).

- *Implementation plan:* If the solution that has been selected requires additional funding, consider all options such as increasing taxes, consolidating or relocating existing stations, charging fees for service, or establishing mutual aid with an adjoining jurisdiction. In addition, a plan that contains phases is very useful to the decision maker. The administra-

tor can then focus on obtaining resources for part of the plan each year, rather than dealing with the entire financial impact in the first year. For example, the plan may call for the construction and staffing of three new fire rescue stations. The administration may be able to use its skills in negotiation, compromise, and political sensitivity to gain approval to construct one new fire station every 2 years. This would implement the plan in 6 years.

Note: If the administration is not successful in gaining approval or financing for the plan or change initiative, the plan should be placed in a top desk drawer for use if an opportunity surfaces. Pull it out next year and try again. Look for opportunities to justify implementing the plan. For example, if the plan is for new fire stations, and a fire death occurs in an area that has an excessive response time, the chief may want to mention this to the administration and elected officials. Allow them to take credit for finding a solution to the fire death.

Future Planning

The concept of future planning in the fire and emergency services organization has been encouraged through documents such as *Urban Guide for Fire Prevention & Control Master Planning* and several courses at the National Fire Academy (Executive Planning and Strategic Analysis of Community Risk Reduction). The following are several items that you may need to consider when using future planning:

- If a project in question is relatively simple, it may be possible to anticipate needs and forestall problems through advanced planning. Many of the short- and long-term planning functions in a fire and emergency services organization are relatively simple. For example, the basic measure of emergency service is response time. The determination of the appropriate response time can be complex, but once the maximum acceptable time is agreed upon, it is a fairly straightforward analysis to determine coverage areas for existing and proposed stations.

- Good reasons exist to base policy on a statement of intent issued by a professional administrator. Politicians and/or senior administration officials would then want to declare to the public that with their support of the new policy or increased level of service, the department can now provide the best emergency service for the community, and thus claim credit for this excellent emergency service.

- If the policy area being studied makes it unwise to delegate broad discretionary authority to the frontline workers, then the major policy decisions must be made at the department level so that each unit provides consistent, reliable emergency service to the public. To allow each company discretion in determining objectives and goals is to invite freelancing in public policy.

- If the implementation effort fits unambiguously within the responsibilities of an existing agency, the appropriate organization to be used for the job will not become the subject of discussion. This can be a problem in some cases, such as when a fire agency proposes to absorb an EMS transport or a consolidation is proposed. Normally this is not an issue, especially when the proposal is for expanding existing emergency services or areas of coverage.

- In the case of fire and emergency services operations, it is not a lack of knowledgeable experts at the company level that is a problem; it is a lack of adequate time and the big picture policy analysis that is needed to weigh all the pros and cons to make a valid decision. That is why standard operating procedures (SOPs) are extremely important in creating a safe working environment and attaining quality, consistent service. The majority of emergency actions by the first arriving companies need to be scripted to guarantee the same results each time. These decisions are best accomplished by a task force process or by adopting an SOP that works from another fire and emergency services organization.

- If there is a high level of confidence in the causal theory underlying the policy, planners can predict the effects of executing changes even in the absence of additional knowledge. Because emergency services are very labor intensive, the administration can predict changes in level of service (e.g., response time) if new people, apparatus, or facilities are added to (or subtracted from) the existing agency.

In many cases, the plan can be implemented less painfully by incremental adjustment. This can spread out a large financial impact over several years, allowing for small increases in tax revenues each year. Politicians usually prefer this approach. Also, as previously noted, this can allow for adjustments to the plan if the outcomes are different than anticipated.

Planners who neglect to consider the desires and special problems of interested individuals in the locale of intended effect are apt also to forget that affected individuals can organize coalitions to resist programs that they judge to be harmful or intrusive. Activists can lobby for legislative changes, can litigate in the courts, can appeal to public opinion, and can manipulate formal procedures in the administrative process. In short, they can often tie up a program that they don't like—almost indefinitely. (Garvey, 1996, p. 466)

One way to overcome this type of struggle is to provide an implementation plan that takes several years to complete and get the elected officials to adopt the plan. Another method that works quite well is to include those individuals who are most interested in and impacted by the planning process.

An Example: Staffing Policy

In relation to the issue of staffing, both OSHA and NFPA agree that a minimum of four fire fighters must be at the scene of a structural fire before an interior attack can be launched. Therefore, one measure of an adequate level of emergency service has to be the time needed for four trained fire fighters to arrive on the scene.

It is difficult also to obtain effective teamwork and coordination with under strength crews. Some fire departments have attempted to solve this problem by supplementing their crews with part-time or volunteer fire fighters, or by providing off-duty fire fighters with tone-activated radio receivers and paying them for overtime when they respond to a fire. The on-duty personnel make the initial fire attack and holding action while off-duty personnel provide the additional assistance needed for continuing fire-fighting operations. Efficiency is lost, and increased fire losses can be expected with this arrangement. Such protection should not be relied upon to replace adequately required staffing and equipment needed immediately at the scene for initial attack and rescue.

Personnel requirements are not merely a matter of numerical strength, but are also based on the establishment of a well-trained and coordinated team necessary to utilize complicated and specialized equipment under the stress of emergency conditions. Attempting to operate more fire companies than can be effectively staffed, even if some response distances must be somewhat increased, is less desirable than fewer but appropriately staffed companies. The effectiveness of pumper companies must be measured by their ability to get required hose streams into service quickly and efficiently. NFPA 1410, *Standard on Training for Initial Fire Attack*, should be used as a guide in measuring this ability. Seriously understaffed fire companies generally are limited to the use of small hose streams [and outside firefighting] until additional help arrives. This action may be totally ineffective in containing even a small fire and in conducting effective rescue operations.

Consideration must be given also to maintaining an adequate concentration of additional forces to handle multiple alarms at the same fire, while still providing minimum fire protection coverage for the other areas under fire department protection. If available personnel prove adequate for routine fires but inadequate for major emergencies, arrangements should be made to supplement the fire protection coverage by calling back the off-shift personnel [or volunteer or call fire fighters] and by promptly calling nearby fire departments for mutual aid. Off-shift personnel may operate reserve apparatus or relieve or supplement personnel on the fireground. Fire companies not dispatched or utilized on the fire scene should be repositioned throughout the remaining area of the jurisdiction to ensure minimum response times to other alarms. (NFPA, 2003, p. 7-37)

Providing on-duty crews is the key to a fast, coordinated response to an emergency incident. In most cases this means career personnel, but volunteer departments have been able to provide on-duty crews by assigning a duty time to each of their members.

This has worked very successfully in a small number of volunteer departments, and this trend seems to be spreading. In contrast to the prediction of some volunteer officials, morale is quite high in these volunteer on-duty departments, and turnover is low. It appears that the volunteers feel that their department really needs their help, and they respond to this need.

In areas where public funding is not available for career fire fighters and the volunteers are not able to provide complete coverage of the station with on-duty members, staffing must rely on alerting volunteer members by outside sirens or pagers to respond to the emergency incident. Generally, these volunteer departments are in areas that serve 2,500 or fewer residents.

Policy Analysis Reference Sources

One of the most comprehensive outlines of public fire and emergency services is contained in NFPA 1201, *Standard for Providing Emergency Services to the Public*. This document also references the following sources, which should all be reviewed and incorporated into any policy analysis:

- NFPA 1221, *Standard for the Installation, Maintenance, and Use of Emergency Services Communications Systems*
- NFPA 1500, *Standard on Fire Department Occupational Safety and Health Program*
- NFPA 1561, *Standard on Emergency Services Incident Management System*
- NFPA 1710, *Standard for the Organization and Deployment of Fire Suppression Operations, Emergency Medical Operations, and Special Operations to the Public by Career Fire Departments*
- NFPA 1720, *Standard for the Organization and Deployment of Fire Suppression Operations, Emergency Medical Operations, and Special Operations to the Public by Volunteer Fire Departments*
- Insurance Service Office, *Public Protection Classification Service*; *Grading Schedule for Municipal Fire Protection*; contact the ISO at ISO Customer Service Division, 545 Washington Blvd., Jersey City, NJ 07310-1686.

For the emergency management function, NFPA 1600, *Standard on Disaster/Emergency Management and Business Continuity Programs*, should be consulted. Note that each NFPA standard may contain references to other standards and reference sources that need to be referred to while developing a policy analysis. The following sections are based on the ISO documents and NFPA 1710.

ISO Fire Suppression Grading Schedule

Many appointed and elected officials have used the ISO's *Grading Schedule for Municipal Fire Protection* for public policy analysis of a fire and emergency services organization. Chapter 1 in this text provides an overview of this process and its history.

One very important factor to keep in mind when reviewing a recommendation from the ISO is contained in the following statement from the NFPA *Fire Protection Handbook*: "The purpose [of ISO grading] is to aid in the calculation of fire insurance rates and is not for property loss prevention or life safety purposes" (NFPA, 2003, p. 7-40). Therefore, if the department is looking for recommendations to improve emergency services to the public, use the ISO recommendations as part of the analysis, but not exclusively.

One idiosyncrasy in the ISO grading schedule revolves around a basic measure of service. As previously discussed, one outcome measure of a fire and emergency services department's service is response time. The sooner units are able to arrive at the emergency incident the lower the amount of property loss and the more lives that will be saved. "Nothing is more important than the element of time when an emergency is reported. Fire growth can expand at a rate of many times its volume per minute. Time is the critical factor for the rescue of occupants and the application of extinguishing agent" (NFPA, 2003, p. 7-311).

The ISO measures fire company coverage by using a response distance calculation. If the area served by the station is within $1\frac{1}{2}$ miles of the station, then the department gets full credit for this item. However, using only distance can be misleading and will result in big differences in response time from area to area. For example, a fire station that is located in an urban area that has heavy traffic or a volunteer station that relies on members traveling to the station before responding will have substantial increases in response times.

Some people have questioned the Class 1 rating (the best public fire protection rating given by the ISO) being given to departments that have fewer than four fire fighters per company. They ask how a department with only 50% of the staffing (i.e., 6 fire fighters) for which ISO gives maximum credit can gain the best fire suppression rating.

There are two advantages to using ISO ratings. The first is it is essentially free to the municipality, and what elected official would turn down a free offer for consultants to review the fire services? As in all other areas, there is no free lunch; somebody always pays. In this case, the costs are added to insurance premiums. The other advantage is that the chief can characterize any improvements that would lower the public fire protection rating as being able to reduce insurance costs to the owners of property in the city. This can be a strong argument.

However, to be completely honest, only owners of certain structures receive substantially reduced insurance premiums. Property protected by automatic sprinklers or highly protected risks (large industrial complexes) already have lower rates. Single-family dwellings are also rated in standard policies that will not see lower premiums unless the rating is dropped several classes. For example, in Florida, the insurance for single-family dwellings and structures protected by automatic sprinklers is essentially the same for Classes 1–7. In addition, several national jurisdictions with populations of over 200,000 have their insurance rates based on loss experience, with the ISO rating having no effect.

One reason to use the ISO grading shows up in its data, which does show a direct correlation between a better classification and less fire loss. "ISO reviewed the cost of fire claims per thousand dollars worth of insured property by PPC for communities around the country. . . . based on five years of data for homeowners and commercial property insurance—show that the communities with better classifications experienced noticeably lower fire losses than communities with poorer classifications" (ISO, 2001, p. 3).

Because these ratings are very sensitive to reduced response distances, adequate staffing, training, and water supply, it is not surprising that departments that have these resources are able to be more effective at extinguishing fires.

ISO and State Farm Insurance State Farm Insurance has stopped using ISO grading in its calculations for insurance covering single-family dwellings. This company is the largest writer of home insurance, covering over 15 million homes. State Farm reports, "By using its own loss experience for all incurred causes of loss as a tool in setting local rates, State Farm will be able to reduce operating costs, improve our ability to respond to actual claim cost trends with a particular area, and ultimately save money for our customers. . . . Since

70% of State Farm's Homeowners Program claim dollars nationally go to pay for non-fire losses, it is appropriate to consider local geographic definitions besides Fire Protection Area in rating plan design" (State Farm, 2001, p. 1).

Insurance companies that use the ISO ratings have to pay a fee for this service. This fee is ultimately paid by the customer, through higher insurance rates. State Farm is able to offer its insurance at a lower price because it does not have to pay for this service.

ISO is actively promoting its services to offset this trend. If all fire insurance were priced using actual experience, then there would be no need for ISO ratings. State Farm notes "The quality, and more importantly, the effectiveness of local fire service clearly have an impact on this claim history" (State Farm, 2001, p. 1). However, many fire officials have relied on ISO and typically use its recommendations to help justify improvements in their budgets.

One problem with using ISO for policy analysis, especially in some cost-benefit calculations, becomes evident when the financial pluses and minuses are considered. Any improvements for the department will be funded by tax dollars. Any benefit to the citizens and property owners comes from lower insurance premiums. To the average person, the two may not seem connected or noticeable. For example, a business owner may pay less insurance but have to pay higher property taxes to fund improvements in public fire protection.

It is incumbent upon the department to educate the business community and elected officials about the advantages of a more effective fire and emergency services department. Many jurisdictions have used lower insurance premiums to attract new commercial development and, as can be seen from the fire loss data from ISO, complying with its recommendations does make fire departments more effective. This also would be true for any jurisdiction using NFPA 1710.

NFPA 1710, *Standard for the Organization and Deployment of Fire Suppression Operations, Emergency Medical Operations, and Special Operations to the Public by Career Fire Departments* This standard was first issued by the NFPA in 2001, so it is relatively new to the fire service. There was opposition to the adoption from many special interest groups, and this opposition continues. For example, the following article from the International City/County Management Association (ICMA) sums up their position:

NFPA 1710: Your Community May Be Next

In recent weeks, several managers have informed ICMA that standard 1710 of the National Fire Protection Association is being specifically cited in their communities as a means to leverage more resources *for* only one aspect of the fire service: staffing and deployment.

Other important priorities, including fire prevention efforts, police protection, general emergency preparedness, and other services of local government are being negatively affected as a consequence. The managers in these communities are asking *for* assistance.

Proponents of increased fire staffing state that NFPA 1710 requires a minimum of *four* persons in each fire company and there are no exceptions. This is not accurate and is an incomplete citation of 1710. At the beginning of 1710, the NFPA Standards Council included an equivalency clause stating: "Nothing in this standard is intended to prohibit the use of systems, methods, or approaches of equivalent or superior performance to those prescribed in this standard. Technical documentation shall be submitted to the authority having jurisdiction to demonstrate equivalency."

By inserting this equivalency clause into 1710, it is clear that the standards council determined there are other equivalent approaches to fulfilling 1710 than just the input-type measures (e.g., minimum four-person staffing, response times cited elsewhere in the standard).

The International Association of Fire Chiefs (IAFC) notes in its *1710 Decision Guide* that equivalency must be based on reasonable and sound principles that could be explained to a jury. IAFC's *1710 Decision Guide* goes on to state that the analysis of resource deployment in relation to risk, which is a component of the voluntary accreditation process of the Commission on Fire Accreditation International (CFAI), is an option to consider in order to meet the equivalency clause.

Another related approach would be to use the *Standards of Response Cover* document being developed by CFAI and expected to be released in December 2002. [This document will be made available free-of-charge on the Web site of the Public Entity Risk Institute.] Clearly, there are more options for equivalency that could be developed by managers, fire chiefs, and corporation counsels and then presented to elected officials for their consideration and action.

Another significantly different approach has been taken by at least one community. This community reviewed the process by which 1710 was promulgated and concluded that the process was biased, special interests were predominated, and that the process was not adequately supported by relevant scientific studies. The council of that community then passed a resolution indicating it did not recognize the validity of NFPA 1710. The factual information to support such a resolution is contained in the appeal presented to the NFPA Standards Council by the Local Government Coalition.

Individuals with examples of how NFPA 1710 is affecting their communities or who have other ways of addressing this issue are asked to send an email to mlawson@icma.org.

Finally, although 1710 is an approved standard of NFPA, proposals for revising or changing 1710 are being considered and must be submitted to NFPA by January 3, 2003. More information on this will be provided in a future newsletter.

Without getting into specifics, some of the items in this article are either misleading or inaccurate. Remember, NFPA 1710 is a new document, and individuals' first impressions may differ depending on the person or organization's viewpoint. As is true with any document, the committee may not have been as clear in its intent and misunderstandings are very possible.

Both the International Association of Fire Chiefs (IAFC) and the International Association of Fire Fighters (IAFF) originally published separate documents in 2001 explaining their interpretation of this standard. Not surprisingly, there were some differences. However, in a newer analysis of this standard by these two groups, they have come to agreement and have jointly published a new document titled *NFPA 1710 Implementation Guide* (2002). This is an excellent source for any analysis on staffing and deployment of fire and rescue services. Following are a few excerpts that help clear up some of the more controversial issues:

The International Association of Fire Fighters and the International Association of Fire Chiefs jointly developed this *NFPA 1710 Implementation Guide* to assist labor and management in working together to take fire and emergency services to a higher level in their communities.

NFPA 1710 establishes a quantifiable method of measuring the quality of a fire department—and in this business quality is defined by the ability to save lives and property. It sets adequate and appropriate guidelines for staffing, response times, and other factors vital to the performance of a fire and EMS department's duties. And in those communities that implement this new international standard, NFPA 1710 will surely save lives of citizens and fire fighters.

Every fire service leader and every local government official should enthusiastically support implementation of NFPA 1710. The benefits of NFPA 1710 compliance, including reduced property loss, far outweigh the arguments of those critics who suggest that the benchmarks in NFPA 1710 are unattainable or too costly for their community.

The goal of this implementation guide is to give fire chiefs, fire fighter locals, and city officials the knowledge, the data, the tools, and a step-by-step process to evaluate their fire and EMS departments and work toward compliance with NFPA 1710. (IAFC & IAFF, 2002, cover letter)

Why We Need Standards

Fire growth and behavior are scientifically measurable, as are the expected outcomes associated with untreated cardiac arrest, and the specific resource requirements to control fires and to prevent deaths. Despite these facts, many communities approach fire/rescue organization and deployment as if it were all art and no science—and abstract art, at that.

The NFPA 1710 standard could be found highly relevant to the question of whether a jurisdiction has negligently failed to provide adequate fire or emergency medical protection to an individual harmed in a fire or medical emergency. To prevail in such a claim, the individual would have to show that the jurisdiction failed to provide the level of service required by the standard, and that this failure was a cause of his or her injury. (IAFF & IAFC, 2002, p. 1-2)

TABLE 11-2 is a comparison between the requirements of NFPA 1710 and the requirements in ISO (to gain the greatest credit in the rating calculations). Although communities getting the best rating of Class 1 do not need to maximize each category, this comparison gives some idea of how the two differ.

In general, a department using NFPA 1710 for deployment and staffing will have a lower budget than a Class 1 ISO department because fewer stations are needed to meet the response times (distances) and fewer fire fighters are needed per company. The calculations for costs using ISO are very complex, so it is hard to generalize for other protection classifications, although some fire experts have commented that they think the break-even point in a cost-benefit analysis for an ISO rating is somewhere around a Class 4 rating.

Table 11-2 NFPA 1710 vs. ISO Class 1 Requirements

	NFPA 1710	ISO
Engine company travel distance	2–2.5 miles*	1.5 miles
Staffing	4 fire fighters	6 fire fighters

*Note: Because NFPA 1710 uses response times, the distance will vary depending on items such as road layout and traffic conditions.

Changing Social Perspective

Administrators conducting policy analyses should become aware of changes in the attitudes and values of new members. Many new members believe they have a right to participate in the decision-making and planning process. Management techniques such as total quality management (TQM) have given the impression that all decisions will be made in small committees of members in some sort of democratic process. Formal task groups in many organizations are slowly replacing the *good old boys* network that was traditionally used for important decision making and controlling the organization.

Both the customers and members of the department are more mobile, on the average changing their residence every 5 years. Most fire and emergency services members used to live in the community they served. This is no longer true; many members, both volunteer and paid, live many miles away from the districts they serve. Lengthy commutes do not seem to bother these members, especially when working only 1 out of every 3 or 4 days.

This presents two problems for the fire and emergency services organization. The first is that members can be less committed to providing public safety service in a town they do not call their own. The second is that the longer commute can present a problem for those departments that rely on callback or volunteers to cover large-scale incidents. Some fire and emergency services organizations have adopted residency or maximum commuting distance requirements because of these unwanted effects. This problem pertains more to small departments, but it can have some impact on larger departments as well.

The typical American family has undergone dramatic change. Studies indicate that a family has the most impact on the values of its members. Over 24% of children now live in single parent families. Almost half of American marriages end in divorce. Fire and emergency services members must cope with complex demands created by the modern dynamic family, psychological trauma caused by family problems and situations, single parent issues, and paternity/maternity absences. Many of these problems end up at work.

Different Generations

The members of the administration and the organization can affect policy analysis. Therefore, it is important to know the characteristics of the different generations. Baby boomers, born between 1946 and 1965, typically reflect values of a strong work ethic, believe in the integrity of family, are more trusting of government, and show more loyalty to their employers. Those born after 1970 may accept divorce more readily, tend to distrust government, and change jobs more frequently.

Several conflicts of values can be seen in these two groups; for example, group commitments versus individual needs, respect for authority and obedience versus participation and democratic rule, a melting pot of cultures and races versus diversity and appreciation for differences, and materialism versus value gained through new approaches and experimentation.

More jobs are requiring a college education, including fire and emergency services organizations. For example, to become a New York City fire fighter, an applicant must have completed 30 college credit hours.

Most new members have no military experience or trade background. Discipline must be instilled in all new members because they no longer have this prior exposure from military service. Many new members also lack some basic skills with hand tools such as a screwdriver or hammer. In previous times, it was typical to expect that all new members had experience with some basic hand tools.

Empowering Employees

Empowerment has been praised by many management gurus as the way to put policy analysis at the worker's level. "It's become a politically correct mantra. . . . Empower, empower, empower. I ask people what they mean by that and they either become inarticulate or they look at me like I'm an idiot" (Kotter, 1996, p. 1001). Like anything else from the business world, adopting and implementing the latest management miracles must not be done without detailed study.

In a manufacturing situation, where each section has to improve its productivity, it would be good to empower the first-line supervisor to make the decisions necessary to accomplish their goals. Once these supervisors understand the vision, those that are closest to the actual work being done can facilitate the smooth operations and any needed changes. For example, workers noted that on numerous occasions they were waiting for one specific part to complete a product. A first-line supervisor may use ingenuity and creativity to find a new fast source for the part. Or, the supervisor may find that assigning additional workers to one phase of the process will speed things up.

This empowerment could cause havoc if used by front-line emergency service workers at the company level. Would it be good for paramedics to choose the medical protocol on an ad hoc basis for each call? Or, should the members of each engine company be allowed to change the location and setup of equipment on their engine? What if the fire and emergency services organization had to follow one and only one SOP: Members are to do the best they can do on each emergency call? With no consistency or adherence to generally accepted practices and procedures, each shift or crew or station or battalion would do its own thing. This would be the ultimate in freelancing, which has been consistently identified as a major problem at many challenging emergency incidents.

So if there is a need for policy analysis or change in operations for emergency service, empowerment is not a good idea. However, if, for example, the chief assigns an officer the responsibility to supervise, plan, and create a new collapsed building and confined space rescue training area, this officer may be empowered with the authority to do whatever is necessary to complete the project. Full, unequivocal empowerment may not be legally possible when it comes to spending public funds, however. If empowered, this officer may be able to apply for donations from contractors and other interested parties or to seek out grants to help defray the costs.

Consultants

There are numerous fire and emergency services consultants who can do professional policy analysis and master planning. The following is a list of items to consider before deciding to obtain professional consultant services:

- The first question is: Does the department have members with the skills, knowledge, and talent to do this type of planning in house? In many cases, the professional planners needed to produce a master plan may not be available in the organization or the municipality. If the answer is yes, would the elected and appointed officials (and the public) have trust that the results are not self-serving? Sometimes, for political reasons, it is better to have an outside consultant make the recommendation, especially if there is controversy or the plan will be costly.
- Be a part of the selection process for the consultant. Check out their past work and look

for any signs that the consultant has been the *hired gun* who comes into town with the sole goal of reducing the fire and emergency services' budget. Make phone calls to fire and emergency administrators in organizations that have been studied by the consultant.

- Cooperate fully and be as accessible as possible. Remember, it is not realistic to expect any consultant to not have some constructive criticism of the department. No person or organization is perfect. There is always room for improvement. Use these recommendations to the department's advantage. However, many shortcomings in the department can be solved only with greater funding and an increase in staff.

CASE STUDY #1: Health Care Spending

The argument for health care as a *public good* is very strong. It has so much support that spending for health care in the United States is greater than in any other country in the world. "Health expenditures in the United States grew 7.7 percent in 2003 to $1.7 trillion, down from a 9.3 percent growth rate in 2002. On a per capita basis, health spending increased by $353 to $5,670. Health spending accounted for 15.3 percent of Gross Domestic Product in 2003, outpacing growth in the overall economy by nearly 3 percentage points" (CMS, 2005).

For comparison, a 1998 national survey by staff of the Phoenix (Arizona) Fire Department reported a medium cost per capita for public fire protection of $95.52 and a standard deviation of $36.23. Even for the higher cost per capita at close to $142 per person, the cost for public fire protection is many magnitudes lower than for health care. There are, however, many similarities that make the study of health care helpful in understanding fire and emergency services organizations.

Although the costs for health care are very high, the patient pays only a small fraction of the cost directly. Fire and emergency services organizations typically do not charge any fees for fire emergencies. Many do charge for emergency medical transports, but in most cases these fees are paid by medical insurance.

Bills for healthcare services are passed on to the insurance company, Medicaid, or Medicare for payment. The patient may pay some of the costs such as a co-payment. For the indigent or those without healthcare insurance, the bills go unpaid by the patient and these costs are added to the overhead costs to run the hospital or medical practice. These costs are subsequently passed on to the patients who pay.

Generally the customer is not billed for any of the cost of public fire suppression services. These costs are generally paid up front through tax revenues and in smaller volunteer fire departments donations help by paying an average of 19% of the budgets (NFPA, 2004).

As explained previously, governments may be justified in regulation where there is a market failure for a public good. For example, without private and government medical insurance plans, it is conceivable that many people would not receive medical care from hospitals or doctors. The market fails because there is no incentive for people to save the money needed for a future injury or illness and therefore there would be many people who could not afford the care. Therefore, health care would not be available.

In the fire and emergency services arena, the customer does not generally pay for either the services or the subsequent losses. Usually a fire insurance company pays the direct losses. The exception to this involves many apartment dwellers or renters, who do not have insurance on their apartments' contents.

Another argument that helps justify government involvement in the healthcare industry is called *specific egalitarianism*. This is the idea that certain goods and services should be provided fairly and with equality to all persons. Because federal and state governments heavily regulate the healthcare industry, the services and competencies of hospitals and doctors are fairly consistent throughout the country.

For fire and emergency services it would be easy to make the same arguments for *specific egalitarianism*. This is especially true because the provider (the fire and emergency services organization) has better information about the quality of the service being provided than the customer.

Again, why is there such heavy regulation for the healthcare industry and not for the typical fire and emergency services organization? The answer probably lies in the differences in the *free-rider* problem between the two public goods. There are many reasons why people do not buy healthcare insurance on their own, such as they cannot afford insurance.

Generally, most owners of buildings have fire insurance for their properties, mainly because insurance is required to obtain financing for property purchase or building construction loans. Therefore, it is almost universal that most fire losses are reimbursed by insurance. There is no free-rider problem, and this occurs without government involvement. Therefore, if owners did not have insurance, they wouldn't get paid after a loss and would not be able to pay back any loans. This is economics and business management 101.

There are other similarities. There can be a disincentive problem for prevention and safety created by insurance. ". . . people who are insured are less inclined to take care to prevent illnesses [or accidents], just as a homeowner insured against fire is likely to have fewer smoke detectors. For example, people with medical insurance may be more inclined to engage in unhealthy activities like smoking, drinking, and overeating, and less inclined to undertake healthy pursuits like exercise" (Bruce, 1998, p. 337).

Because fire insurance removes most of the financial consequences for people being careless, being served by ineffectual public fire agencies, or not providing adequate built-in fire protection systems, this becomes a very strong argument for government regulation. In most cases there is no adverse consequence to the individual, so for the safety of the community and its occupants, minimum levels of fire safety have been dictated typically by building and fire prevention codes.

In this country, there are several building and fire codes that essentially provide a minimum standard for fire safety in structures where people live and work. The oldest fire safety code, NFPA 13, *Standard for the Installation of Sprinkler Systems*, is a little over 100 years old.

Valid justification for government regulation requiring automatic sprinklers can be seen in the following examples:

- Buildings without automatic sprinklers could have an adverse economic impact on the community and its residents.
- Fires in these buildings may be beyond the capabilities of the local fire department.
- There is a real concern that occupants would die in a fire before the fire department could arrive.
- When the structure is on fire there is a safety hazard to fire fighters.

Discussion Questions

1. Why is the spending per capita so much higher for health care as compared to fire and emergency services?
2. Why is there more state and federal funding and regulation for health care?
3. Do our customers believe healthcare workers are more competent and professional than fire and emergency services responders? Why?
4. How does the very low cost to the customer affect the demand for these two services?

CASE STUDY #2: The *Quint Concept*

One major metropolitan fire department has adopted a new strategy for equipping and staffing its fire companies called the *quint concept* (this is a fire apparatus that is constructed to replace an engine and ladder company with one piece of equipment containing hose, water, a pump, and an aerial ladder). The technical merits of this decision will not be discussed, only other, more prominent influences that led to this complete transformation. Some experienced fire officers have observed that this type of apparatus should have a crew of six or seven to fully use the quint's potential; however, this department uses four fire fighters for each company. When talking to the senior officials of this department, they are quick to point out several tactical advantages that mainly focus on their ability to direct many elevated streams quickly on major defensive strategy fires. However, the original pressure for the new policy may have come from other sources.

This city was faced with a substantial shortfall of tax revenues, and the elected officials were pressuring the fire department for reductions in personnel. The fire department had also not been able to purchase any new fire apparatus for many years.

Generally, many older central cities in this country are protected by fire departments that have a long tradition and history. These older departments typically have some stations that were constructed very close to each other at a time when horses pulled the steam pumpers. At those times, to reduce response time, stations had to be spaced closer, but this is no longer necessary with motorized apparatus (although in some cities, the traffic is so heavy that even closely spaced fire stations have excessive response times).

In addition, most of the older large cities have a greater number of ladder companies than typically found in newer cities or suburban areas. In the past, there was justification for these ladder companies to be able to rescue trapped occupants from windows in taller buildings. The enforcement of modern building and fire codes has eliminated many of these *fire traps*; many taller buildings have automatic sprinklers and fire-resistive exit stairs. It was also common to have large crews of fire fighters who worked long workweeks at low salaries.

This major metropolitan fire department operated 36 engines and 16 ladders out of 30 stations totaling 52 companies before changing to quints. After the conversion, the department ended up with 30 quints and 4 ladders totaling 34 companies. Slightly over one-third of its fire companies were disbanded. Even with this major cutback in personnel and companies, the department managed to keep all fire stations open by placing one quint engine in each of the 30 stations.

Although good changes can be the outcome of a major cutback in funding, there may be more than one reason that caused this particular change. With this type of forced change there is always the possibility that this new concept may have been fabricated to make the round peg fit into the square hole. It has been argued that the original pressure for this change was not a better way to extinguish building fires, but to comply with a major cutback in the fire department's budget.

Discussion Questions

1. How would you research the facts in this case to assess their accuracy?
2. If the original justification was budget cutting, is this new system actually giving better or equivalent fire protection?
3. What harm would there be from not informing the public of all the circumstances behind the new policy decision?
4. Do our customers believe we are their protectors and trustworthy, and therefore are blindly trusting our policy decisions?
5. How does reduced revenue for the city affect our customers' demand for our services?

CASE STUDY #3: Smoking Ban

Smoking is now widely accepted as being very dangerous to the well-being of individuals. It has been reported that the number of smokers in the general population has now fallen to less than 25%. The demonstrated ill effects of secondhand smoke on nonsmokers have prompted bans on smoking in numerous public buildings (e.g., restaurants, offices, public assemblies). The federal government has now banned smoking on all domestic airplane trips.

Many fire departments provide automatic retirement benefits for fire fighters developing any heart/lung illness. One major cause of these illnesses is smoking cigarettes. An article describing the results of studies by the London School of Hygiene and Tropical Medicine indicated that smoking has greater effect than fire fumes on fire fighter's lungs. The original justification of the *heart/lung benefits* was statistical studies completed in the early 1950s. In today's modern fire departments, however, self-contained breathing apparatus (SCBA) is mandatory and prevents the majority of toxic fumes from being inhaled by fire fighters.

A 1975 study by H. D. Peabody discovered that cigarette smoking and exposure to carbon monoxide interacted in a synergistic fashion to substantially lower pulmonary function. This reduces the ability of the fire fighter to perform the heavy work for sustained periods of time that is necessary during firefighting. Paul O.

Davis, Ph.D., in *Postgraduate Medicine*, August 1982, had the following comments, "In as much as smoking increases risk of coronary heart disease [number one cause of on-the-job fire-fighter deaths], the examining physician should discourage it. Smoking can only aggravate the pulmonary and cardiovascular effects of the toxic gases to which fire fighters are exposed" (Davis, 1982, pp. 41–46).

In March 1978, the city of Alexandria, Virginia, refused to hire smokers as fire fighters because smokers accounted for a disproportionate number of early retirees from the fire department. There seems to be a strong argument to support banning cigarette smoking by all fire fighters. However, there are some practical and legal questions (policy) that need to be resolved when implementing this type of policy.

Many unions or individuals will argue that this is an infringement on their personal rights (e.g., Why can't these fire fighters smoke when they are not on duty?). One could predict an unpleasant legal and political fight before a smoking ban this comprehensive becomes accepted.

Discussion Questions

1. Would it be reasonable to adopt the policy of not hiring smokers and requiring all new fire fighters to sign a condition of employment that they will not smoke while employed as a fire fighter? Annual medical exams could test for nicotine as part of the enforcement process.

2. Should the city fund a recognized program for smokers to quit and encourage fire fighters to attend?

3. Should a policy that prohibits smoking in the fire stations be adopted?

4. Would this new policy save the city/county money by reducing costs for medical insurance, reducing sick leave, and in general providing a more physically fit fire fighter? Provide references.

CASE STUDY #4: Is Rural Public Fire Protection Effective?

The following are excerpts from an article by Larry Davis (2002) called "Dollars & Sense: The Inadequacies of Rural Fire Protection," found in *FireRescue* magazine. Please read these quotes and answer the questions that follow:

"...our 30-year national effort toward reducing America's fire problem haven't impacted the bottom line. The rate of fire fighter deaths and injuries in structure fires hasn't changed..." (Davis, 2002, p. 78).

"In addition to these figures, civilian fire deaths and injuries in the United States still exceed those of any other industrialized country in the world" (Davis, 2002, p. 79).

The following are specific statements by Davis in this article (Davis, 2002, pp. 77–84):

- "Rural communities adopt few building and fire codes because they lack full-time municipal governments."
- "Even where codes exist, they are seldom enforced due to lack of adequately trained personnel."
- "Rural fire departments are usually totally volunteer."
- "Fire stations are generally located in more populated areas, lengthening response times to less densely populated or boundary areas."
- "Neighboring rural communities with sparse populations can offer little or no automatic or mutual aid firefighting resources."
- "Rural communities have the highest rates of fire death and property loss (more than double) in any population class."
- "Seventy-three percent of rural fires occur in residential structures lacking operational smoke detectors."

"In the early 1970's, the International City Management Association (ICMA) conducted an interesting study about the value of rural fire departments regarding property loss. The report, titled *The Critical Fire Problem in Small Towns*, concluded that many communities with populations fewer than 5,000 people would be better off with *no fire department*. When a fire occurs in one of these communities, however, the dollar loss is often the same as if the fire department had never responded." "Based on all of the statistics, most rural fire departments probably are of greater value to the citizens they serve for providing EMS, vehicle extrication, other rescue services and nonemergency assistance than providing fire-suppression services" (Davis, 2002, pp. 77–84).

"Given the statistics, providing every household in the rural community with smoke detector(s), and then checking them once a year to make sure they're operational would save far more lives from fire than all of the fire trucks and fire fighters in the world." ". . . unfortunately, fire fighters join to fight fires and make other emergency responses and don't generally want anything to do with providing the fire safety education their citizens really need" (Davis, 2002, pp. 77–84).

Discussion Questions

1. If the previous excerpts from Davis are factual (at your option, you may acquire your own studies or statistics to refute his findings), would you adopt a policy to accomplish the installation of smoke detectors in all dwellings in a rural community?
2. How would you implement this public policy?
3. Is the substantial increase in property loss and deaths related to the fact that rural volunteer organizations rely on the response of personnel at the time of the incident (longer response times) as compared with career and on-duty volunteer agencies that have crews ready to respond at all times? Does the training (competency) of these fire fighters have any impact on this anomaly? ("An estimated 233,000 firefighters, most of them volunteer serving in communities with less than 2,500 population, are involved in structural firefighting but lack formal training in those duties" [NFPA, 2004, p. v].) Or are only excessive response times to these incidents to blame? (In many cases, response distance to the greatest number of structures is within generally accepted limits because most are in a central town where the fire station is generally located.) Or, is it a combination of causes?
4. How would you respond to the proposed reduction in the funds available for a new pumper to replace a 25-year-old pumper when those funds are being redirected to provide smoke detectors in dwellings?

General Discussion Questions

The following are suggestions and strategies for the fire and emergency services administrator to consider in justifying their budget. There are no right or wrong answers to these questions, but the administrator should answer each question to their satisfaction and gain input from other managers in the organization. These questions assume that the executive official or the elected legislature has cut the department's proposed budget. For this assignment, choose a major program in a fire and emergency services agency, and then write an answer to each of these questions:

1. Should you mount a campaign to support your proposed budget using family, friends, and the fire fighters' union?
2. Should you work quietly within the system to determine the political feasibility of fighting publicly?
3. Should you fight just enough to show your employees that you care, but not enough to do political damage to yourself? (How would you determine what is "just enough"?)
4. Should you leak information to the media about the disastrous consequences of the budget cut?
5. Should you accept the cut and prepare to be creative about how you implement the cut?
6. Should you be prepared to point out problems with the budget cut when the next catastrophic fire or emergency incident occurs?

7. How would the old adage, "You don't want to win the battle and lose the war," affect your actions?

8. What influence would knowing that your boss has the authority to fire you have on your actions?

References

Bruce, Neil. *Public Finance and the American Economy.* Reading, MA: Addison-Wesley, 1998.

CMS (Centers for Medicare & Medicaid Services). *Health Care Spending in the United States Slows for the First Time in Seven Years.* http://www.cms.hhs.gov/media/press/release.asp?Counter=1314 (accessed August 21, 2005), 2005.

Davis, Larry. Dollars & sense: The inadequacies of rural fire protection. *FireRescue Magazine.* November 2002, pp. 77–84.

Davis, Paul O., R. J. Biersner, R. James Barnard, & James Schamadan. Medical evaluation of fire fighters: How fit are they for duty? *Postgraduate Medicine,* 72(2), pp. 241–248, 1982.

Edwards, Steven T. *Fire Service Personnel Management.* Upper Saddle River, NJ: Brady/Prentice Hall Health, 2000.

Garvey, Gerald. *Public Administration: The Profession and the Practice.* New York: St. Martin's Press, 1996.

IAFC & IAFF. *NFPA 1710: A Decision Guide.* Fairfax, VA: IAFC, 2002.

International City/County Management Association (ICMA). *NFPA 1710: Your Community May Be Next.* http://www.icma.org (accessed November 24, 2002), 2002.

ISO. *ISO's PPC Program: Helping to Build Effective Fire-Protection Services.* Jersey City, NJ: ISO Properties, Inc., 2001.

Kotter, John P. *Leading Change.* Boston, MA: Harvard Business School Press, 1996.

Lindblom, C. E. & E. J. Woodhouse. *The Policy-Making Process, 3rd ed.,* Upper Saddle River, NJ: Prentice Hall, 1993.

McElroy, Wendy. *Reading between the Numbers.* http://www.foxnews.com/story/0,2933,114991,00.html (accessed July 24, 2005), 2004.

NFPA. *Fire Protection Handbook,* 19th ed. Quincy, MA: NFPA, 2003.

NFPA. *State-by-State Findings of Fire Department Needs & Response Capability.* Quincy, MA: NFPA, 2004.

Phoenix Fire Department. *1998 Phoenix Fire Department National Survey on Fire Department Operations.* Unpublished, 1998.

Schick, Allen. *Budget Innovation in the States.* Washington, DC: Brookings Institution, 1971.

State Farm. *Frequently Asked Questions about Subzone Rating.* Unpublished paper. Bloomington, IL, 2001.

USFA. *Advanced Fire Administration: Course Guide.* Emmitsburg, MD: USFA, 1998.

USFA. *Fire in the United States 1992–2001, 13th ed.* http://www.usfa.fema.gov/statistics/reports/pubs/fius13th.shtm#accessible (accessed August 21, 2005), 2004.

USFA. *Historical Overview.* http://www.usfa.fema.gov/fatalities/statistics/history.shtm (accessed July 24, 2005), 2005.

Wilson, Paul F. *Root Cause Analysis: A Tool for Total Quality Management*/Paul F. Wilson, Larry D. Dell, and Gaylord F. Anderson. Milwaukee, WI: ASQC Quality Press, 1993.

12

The Future

Knowledge Objectives

- Comprehend the creation, effect, and importance of progressive change.
- Understand the impact and influence that the future will have on fire and emergency services administration.
- Examine and comprehend the effect that outside influences have on the future.
- Identify the responsibility and accountability of administration to constantly strive for improved service to the public.
- Comprehend and understand the ability to predict future trends in the public fire and emergency services.

Persistent Sense of Urgency for Change

The urgency for change in the business community is necessarily of a higher intensity than what is normal in public safety. For individuals in the free enterprise business community, where turning a profit is the number one goal, there is the real possibility that the company might go out of business by not keeping up with the latest and greatest consumer products and services. In most cases, this creates a constant urgency for change.

Most of the urgency to make changes in the fire and emergency services organization must come from within. However, there are still some outside pressures that will require changes, such as regulations including Occupational Safety and Health Administration (OSHA) safety standards. Even with these regulations and standards for operations and safety, in many cases

change must be the result of competent leadership at the local level. That is where the everyday urgency will be created and found.

Measuring Performance

John Kotter (1996, p. 162) concludes that the "[t]ypical employee in typical firms today still receive little data on their performance, the performance of their group or department, and the performance of the firm." This is also true in the fire and emergency services business. Lack of feedback to the members is not the result of poor management, but of a lack of valid performance measurements.

To measure performance in the future, valid outcomes are needed that are the direct result of inputs. A National Fire Protection Association (NFPA) committee adopted a new standard (NFPA 1710) to address some central issues of fire and emergency services: response times and staffing. Now that this is completed, the fire and emergency services community will have legitimate measurements of service delivery and can set goals for the future.

The following advice is accurate and appropriate for fire and emergency services organizations: "The combination of valid data from a number of external sources, broad communications of that information inside an organization, and a willingness to deal honestly with the feedback will go a long way toward squashing complacency. An increased sense of urgency, in turn, will help organizations change more easily and better deal with a rapidly changing environment" (Kotter, 1996, p. 163).

Hiring and Promoting Team Players

The future is in the hands of new members. New positions and promotions should go to those who are team players. This can be a difficult item to assess, but not impossible. There are professional consultants who can help devise tests to screen applicants for acceptable personality traits that would be desirable in a new member. Incumbents who aspire for promotions can be evaluated periodically based on a set of known qualities for team players.

Psychological testing is becoming more and more common in the public safety professions. In 1996 an armed fire fighter came to a Jackson, Mississippi, department's headquarters and killed four chief officers. This is an extraordinary case, but the point is that some individuals are not compatible with the intimate small family atmosphere of the typical fire and emergency services company, which requires a team player and a person with very good interpersonal skills.

In the future, we will also be looking for individuals who enjoy and feel comfortable interacting with the public. Previously, we have typically attracted and accepted individuals with very similar personalities who preferred to interact with those with similar personalities. Between diversity issues and the need to interact with the public, hiring people who are tolerant of others and want to help the public in the same way a teacher or nurse helps is preferred.

In the past, the typical new member was a risk taker who was physically strong. These traits are still very important, but compassion, open-mindedness, and being a team player are also equally essential now and in the future.

Broad-Based Empowerment

Again, this is one item that has no place in emergency operations. This business technique works best in companies that are constantly changing, such as high-tech and professional service firms. Not every new management technique belongs or has a future in the fire and emergency services. Historically, many management miracles have not stood up to the test of time.

Most of our services are rather low tech, so we need a high degree of consistency to ensure the customer receives the same quality service on each call. If we allowed independent decisions on the tactics and strategy at the company level, there is no telling what would happen from call to call. This situation would compound itself on large-scale incidents where there are specific expectations of the capabilities and operations of all companies.

For example, in a major metro fire department, each company officer was allowed to determine the in-service training program. In essence, these officers were empowered to devise their own training program after making a personal and localized assessment of the hazards in their first due areas.

One officer in a rural area devised a yearly schedule of drills and did not include any drills on the use of fire department standpipes that are used in high-rise firefighting. In the officer's mind the right decision had been made because there were no high-rise buildings in the first due area. In fact, there were not any fire hydrants in this predominantly rural farm area. What was not taken into consideration was that the officer or any of the members of this company could be transferred to another station that could have many high-rise buildings in the first due area. Also, this company could be dispatched to fill in at another station that would have high-rise buildings in their response area.

Technological Impacts on Fire and Emergency Services

The world's knowledge and technology is growing at an ever-increasing rate. Scientists have predicted that our collective knowledge will double every 5 to 10 years.

Many changes in the fire service have been relatively recent. For example, the federal fire focus of the U.S. Fire Administration and the National Fire Academy (NFA) is only slightly more than 20 years old. The first NFPA professional qualification, Standard 1001, *Fire Fighter*, was adopted in 1974. Taking into consideration that the U.S. fire service is over 350 years old, these are recent events. Many of these NFPA fire service standards are still finding their way into the training and operations of the U.S. fire service. Change takes time, vision, and leadership.

Technology seems to be a never-ending source of change, some good, some bad, and some not necessary. Some examples of new technology are 1³⁄₄- and 2-inch hose, positive pressure ventilation, automatic defibrillators, the national incident management system, compressed air foam, infrared cameras, and GPS navigation (**FIGURE 12-1**).

In most cases it is easier to sell change that is technology based as compared with change directly impacting the members of the organization. It can be clearly demonstrated that additional training, especially in the basics, and increases in physical fitness will dramatically improve performance at an emergency incident. Remember that competency at the emergency scene is the result of three separate functions—competent personnel, equipment, and command (SOPs). To have fire and emergency services organizations equipped with the newest and best technology is only one part of the system to provide quality emergency services to the public.

Many of these new technologies may be very helpful for the fire and emergency services organization, but our process for research, experimentation, and acceptance tends to be very sporadic and decentralized. The U.S. fire service is made up of a large number of departments (over 30,000) that characteristically serve small numbers of people and have minimal resources. According to NFPA statistics, there are only 14 fire departments that protect 1 million or more persons. On

the other end of the spectrum, there are 14,817 fire departments that protect fewer than 2,500 people each.

There is no large well-funded "Independent Fire and Emergency Services Research Agency" that can investigate and study new technologies, methods, or equipment. This type of research should be done before the new tools or techniques are used in the field. Today, we learn about tools' effectiveness from other organizations' use of the new technology or from the manufacturers.

Progressive Organizations

There appears to be a bias in some organizations that label themselves as *progressive, state-of-the-art* agencies in their blind acceptance of anything new (predominately equipment). Some fire and emergency services departments have been described by their members and the public in glowing terms because the department was equipped with newer emergency response vehicles. However, some of the best fire and emergency services organizations can have older equipment if staffed by highly competent personnel. Highly competent assumes two items—quality training made up of a thorough recruit training and an ongoing intensive in-service training coupled with personnel who are in good physical fitness.

Just because equipment or a technique is new does not mean it is better, and just because equipment or a technique is old does not automatically mean it is bad. Each new tool, idea, or technique needs to be judged on its own merits using research techniques that are based on standards and broad-based beta testing.

Also, question very carefully information, observations, and studies that have been supported by the funding or equipment from a manufacturer. Not all manufacturers are unscrupulous, but it is unlikely that a balanced study process would be found that leads to a conclusion that would be completely unbiased. It is the manufacturer's job to tell the customer the best reasons to buy its product.

This *newer is better* bias can be found when considering adopting new management philosophies such as management by objectives (MBO), total quality management (TQM), customer service, and many others. Some of these management practices have been used in businesses and general government agencies—some successfully, some not. In many cases, the final results have not been apparent for many years after the first implementation.

FIGURE 12-1 Firefighter Using an Infrared Camera

The future is always difficult if not impossible to predict. Be cautious and take the time necessary to gain as much information as is needed to make your decision. However, 100% certainty is not realistic, so do not let some uncertainty stop the decision and implementation of a new change. Be prepared to adjust if the results are not the expected outcomes.

America at Risk

This study, which was concluded before the terrorist attacks of September 11, 2001, was the first time the concept of total risk management was discussed at the federal level. The future was going ". . . beyond the boundaries of the fire risk alone. *America Burning* in 1973 had anticipated some of the forthcoming challenges to the fire services. . . . The establishment of FEMA, the growth of the emergency management community as a profession, the increase in disaster losses in America, and other factors, had dictated a different context for the fire service" (FEMA, 2002, p. 6). The fire service is on this journey today into the future and the expectations of elected officials and the public have changed.

Another theme that is present in this report is the realization that prevention efforts are very effective. Once a fire has started or a terrorist has activated a bomb, the fire and emergency services are in a reaction mode to the destruction that has already occurred. Because there has not been a major terrorist event in the United States since September 11th, one could assume that extensive prevention efforts are responsible. At this time, the federal government is not talking—they just confirm that security is being pursued through prevention. The following is a list of the major topics discussed in this report:

- Implementation of loss prevention strategies (multi-hazard)
- The application and use of sprinkler technology
- Loss prevention education for the public
- The acquisition and analysis of data
- Improvements through research
- Codes and standards for fire loss reduction in the built environment
- Public education and awareness
- National accrediting and certification
- Fire-fighter health and safety
- Emergency medical services
- Diversity

Higher Education

The results of an informal survey on Firehouse.com indicated that more than 27% of fire chiefs have a bachelor's degree or better. This unscientific study received 7,537 responses. This does indicate that many chiefs have college degrees, but the same study concluded that 32.7% had no college background (Firehouse.com, 2001). The study did not differentiate between the size or type of the departments, but when a municipality recruits for a chief officer, it commonly requires a bachelor's degree as a minimum educational requirement.

At the national level, both the NFA Executive Fire Officer program and the International Association of Fire Chiefs (IAFC) Certified Fire Officer Designation will require a minimum education level of a bachelor's degree in 2009. At the NFPA, the Fire Officer professional standards do not require any college education to comply with their criteria, which are commonly used for certification in many areas. There are some mixed messages to future officers, but to be on the safe side, members who strive to become future leaders in the emergency services profession should pursue a bachelor's degree as a minimum.

One study, *Fire Service Degree Programs in the United States*, reported that there are 222 associate-level programs and 26 bachelor programs. Many serve only the geographic areas near their campuses; however, there is a growing trend to offer courses at some higher education institutions via the Internet. This report went on to comment ". . . the most important reported challenges facing degree programs included means and methods to increase enrollment, updating curriculum, finding quality instructors, lack of funding, and lack of incentives for earning an advanced degree" (Sturtevant, 2001, p. v).

Programs that are specific to emergency services and can be obtained by distance education are good choices, such as the Fire and Emergency Services bachelor's degree program at the University of Florida or the Degrees at a Distance program sponsored by the NFA (at seven different colleges). These bachelor's degree programs all build on the student having an associate's degree before entering the college or university.

One quick observation: Even though many current officers do not have bachelor's degrees, the key to leadership and change management is a person who has higher education. When dealing with elected officials and the public, including union presidents, there is no substitute for education in giving a chief officer the confidence to be successful when leading an organization through change. And without change, there cannot be progress.

Professional Status: The Future of Fire Service Training and Education

The following is a discussion based on a series of articles written in 2004 by Dr. Denis Onieal, Superintendent, National Fire Academy. He attempted to summarize the existing situation and suggest changes to improve training and education in the fire service. He poses the following question by a young person: "Mom/Dad, I want to be a fire chief. What do I need to do?"

If this question had been, "What do I need to do to become a physician?" then the answer is fairly straightforward: 4 years of college, 4 years of medical school, internship, residency, pass the medical boards. For emergency services officers to be accepted as professionals, their preparation must be based on a defined education, training, testing, and experience. "Their [professional] systems of acquiring knowledge are reciprocal among all states. When physicians or lawyers or nurses move from state to state, they may have to present their credentials to the professional board in their new state. They may have to take an exam, or perhaps take some refresher courses—but they don't have to go back to school to learn the basics all over again" (Onieal, 2004, p. 2).

The present situation in the fire service is explained by Dr. Onieal:

It varies from place to place, depending upon the organization, the structure of the department and the governing agency. The process isn't the same wherever you go; frequently it is a slow and uneven process, or one solely based on popularity. Too often the process frustrates talented men and women; we lose our best and brightest. These are the very people who epitomize the word "professional"—the ones who have the aptitude and drive to help the department face new challenges. . . . Right now, there is no one universally recognized and reciprocal system to acquire the knowledge and skills required in the Fire and Emergency Services. (Onieal, 2004, p. 1)

Many of the structural parts for a professional fire and emergency services training and education system already exist. For example, state training agencies, most larger fire departments, and other regional training centers are already in place and providing fire fighter and officer training. The problem is that most have been developed independently and many have different requirements. Even though most use the NFPA professional qualifications, there are differences in the length of these training programs, and testing for certification varies for each state. Clearly, this is one area of emergency service that needs work and some type of consensus in the future. Ultimately, this consensus may be driven by the need for all fire and emergency services agencies to work side by side at major terrorist events.

Some states maintain separate educational systems that provide officer and other specialized fire training. For example, a community college may have a fire science program, but the courses are not recognized by the state training and certification agency. To become a certified fire inspector, applicants must take the courses offered by the state training agency, even when the fire science program in the same state has courses that cover the same educational material.

One example of a seamless education certification system was created in Florida. All the courses offered in the associate's degree fire science programs are the same courses required to take the state certification exams. When students finish their fire science degree, they have also completed the required education to take the state exams for fire officer and fire safety inspector. This does not guarantee success on the certification test, but the reference materials for the tests are generally the same texts that are used for the college courses.

A model education, training, and certification system exists today for the EMS component of emergency services. This system was originally created at the national level by the National Highway Traffic Safety Administration. A national certification system, the National Registry of Emergency Medical Technicians, is accepted by most states. This system is used by 90% of all states, and the other 10% use the same training and skills, but do their own testing.

Homeland Security

The war on terrorism seems to be driving contemporary changes in fire and emergency services administration and operations. On March 1, 2004, the U.S. Department of Homeland Security (DHS) adopted a National Incident Management System (NIMS). Previously, the emergency services had several incident command systems to choose from, each a little different from the others. "This system provides a consistent nationwide template to enable Federal, State, local, and tribal governments and private-sector and nongovernmental organizations to work together effectively and efficiently to prepare for, prevent, respond to, and recover from domestic incidents, regardless of cause, size, or complexity, including acts of catastrophic terrorism" (DHS, 2004, p. ix).

The IAFC has suggested:

- Every responding agency (local, state, national) must use a standardized incident command system (ICS).
- Continual coordination between local, state, and national responders is essential for success in protecting our personnel and the citizens we serve.
- National agencies need to involve local first responders in every level of discussions for resource allocation, especially training and equipment. (IAFC, 2001, p. 3)

Many of the details of the NIMS are scheduled for study and adoption at a later date. This will be an ongoing project for the DHS. For example, specific typing of resources and personnel qualifications have not been developed at this time. These are very complex subjects, especially for the fire service, which has traditionally developed standards of service at the local level. Consistency is rare.

As this project proceeds, expect to see efforts to define and implement the following for responses to a federal declared emergency:

- Mutual and/or automatic aid agreements
- Credentialing emergency responders for different competencies
- Resource typing
- Training for NIMS
- Integrated communications systems

For a look into the future, the federal system for responses to wildfires is a good model. This system is able to provide direction and coordination for hundreds of resources and thousands of fire fighters operating at major wildfires. Wildfire resources and fire fighters are brought in from hundreds and even thousands of miles away and from numerous individual departments and are able to work smoothly side by side. This type of effort to provide efficient emergency services will be needed at the next major catastrophic hurricane, earthquake, wildfire, building fire, and terrorist event.

Who Are We Today?

To see where to go in the future, a person must know where they have been and where they are today. This entire book up until this chapter has discussed where we have been. However, one final look at who we are as individuals in the fire and emergency services will finish this story. There is no better way to know who we are than to see how others see us. To support that perspective, the following article is offered:

The following was stated by a WDVE Pittsburgh DJ on March 15, 2004. WDVE is a hugely popular classic rock station in Pittsburgh, and Scott Paulsen is the morning DJ.

They Don't Make Many

March 15, 2004

Scott Paulsen

I'd like to dedicate tonight's show to the families of Richard Stefanakis and Charles Brace, Pittsburgh firefighters.

Chances are, no one died where you work today. Nobody placed their life in your hands. Odds are that there wasn't a single moment today where you were expected to make a life and death choice at the office, in the shop, or at the store where you work.

And when you left the house this morning, you probably did not think to yourself, this could be my last day.

And if that is indeed the case, good for you. And good for me, as well. Nobody faced death today at this radio station. Not a single one of us arrived at work, knowing that getting your requested Led Zeppelin and that Mullet Talk sketch on the air would save your life. And nobody I work with, to my knowledge, rolled out of bed this morning with the premonition that today could be his or her last.

I don't know that I could work a job that gave me those kinds of thoughts. The worst thing that can happen at my job is that I'm not entertaining, I don't play the right music, people stop listening, I get fired and have to find work at an oldies station that pays one third of what I'm making.

I'm not exactly living in danger. Nor do I wish to.

Thank goodness, not everybody is like me.

There are those people in this world who like to be challenged. They enjoy doing the jobs that no one else will do. These people are motivated by more than just money. They have an inner motivation that gravitates them towards their current job. Ask any one of them why they do it and they'll tell you they can't imagine doing anything else.

They are firefighters.

Each day, firefighters report to their jobs, punch the clock and settle in for what could very well be their last day on Earth. Monday night, Thursday afternoon, Saturday morning, it's the same. The alarm will sound. They will respond. And every once in a while, they die while fighting a fire.

This past Saturday morning, the city of Pittsburgh lost two such firefighters. Richard Stefanakis and Charles Brace had been firefighters for more than 30 years each. That's thirty years of leaving for work, kissing the wife and kids goodbye, knowing that today might be the day. That's thirty years of watching as comrades and buddies fell by the wayside. That's thirty years of answering the bell.

Richard and Charles were inside the famous Ebenezer Baptist Church, the enormous stone structure in the Hill district of Pittsburgh, during the tail end of a multiple alarm blaze that destroyed the 72-year-old building.

There was nothing out of the ordinary about the fire. Just another day at work. While others were preparing themselves to march in the St. Patrick's Day Parade, Stefanakis and Brace, 30 year firefighting veterans, moved hoses throughout what was left of the gigantic Hill District church, trying to make sure there would be no more flare ups. The fire was just about out. But you can never be too careful. They had been battling the blaze since before dawn.

At just past noon, the enormous five-story bell tower collapsed, crushing the two firefighters, throwing block, brick and granite down on the heads of dozens of still-working comrades.

Twenty-eight were taken to hospitals.

Richard Stefanakis and Charles Brace died at the scene.

Just another day at work.

I don't know whether I could run into a burning building. Certainly, you would think I'd be able to do it to save a loved one, my child, my mother, my wife but would I be willing to venture into the depths of Hades to rescue a total stranger? Would I be willing to put my life on the line, just to see if there was someone in that building? And even if I was willing, would my body be able to survive? Would my instincts be enough to get the job done? And, having done the job, would I go back in again. And again. And again.

Week after week after week, fire after life-threatening fire, for thirty years? . . . **No.**

That's what fire fighters do.

Some Good Advice

Investor's Business Daily has spent years analyzing leaders and successful people in all walks of life. Most have 10 traits that, when combined, can turn dreams and high goals into reality in the future:

1. *How you think is everything:* Always be positive. Think success, not failure. Beware of a negative environment.
2. *Decide upon your true dreams and goals:* Write down your specific goals and develop a plan to reach them.
3. *Take action:* Goals are nothing without action. Don't be afraid to get started. Just do it.
4. *Never stop learning:* Go back to school or read books. Get training and acquire skills.
5. *Be persistent and work hard:* Success is a marathon, not a sprint. Never give up.
6. *Learn to analyze details:* Get all the facts, all the input. Learn from your mistakes.
7. *Focus your time and money:* Don't let other people or things distract you.
8. *Don't be afraid to innovate; be different:* Following the herd is a sure way to mediocrity.
9. *Deal and communicate with people effectively:* No person is an island. Learn to understand and motivate others.
10. *Be honest and dependable; take responsibility:* Otherwise, Nos. 1–9 won't matter.

A Final Piece of Advice

Risk Taking Is Free

To laugh is to risk appearing the fool.

To weep is to risk appearing sentimental.

To reach out for another is to risk involvement.

To expose feeling is to risk exposing your true self.

To place your ideas, your dreams before the crowd is to risk their loss and public ridicule.

To love is to risk not being loved in return.

To live is to risk dying.

To hope is to risk failure.

But risk must be taken, because the greatest hazard in life is to risk nothing.

The person who risks nothing, does nothing, has nothing and is nothing.

He may avoid suffering and sorrow, but he simply cannot learn, feel, change, grow, love, or live.

Chained by his certitudes, he is a slave, he has forfeited freedom.

Only a person who risks . . . is free.

—Author Unknown

To promote progressive change in the future, an administrator must have the courage to be a risk taker. The other trait that is absolutely necessary is perseverance. Never give up the fight.

References

DHS (U.S. Department of Homeland Security). *National Incident Management System.* http://www.dhs.gov/interweb/assetlibrary/NIMS-90-web.pdf (accessed August 13, 2005), 2004.

FEMA (Federal Emergency Management Agency). *America at Risk, America Burning Recommissioned.* http://www.usfa.fema.gov/applications/publications (accessed September 27, 2005), 2002.

Firehouse.com. *What Is the Formal Educational Level of Your Fire Chief?* http://www.firehouse.com/polls/2001/ (accessed September 27, 2005), January 15, 2001.

IAFC. *International Association of Fire Chiefs. Leading the Way: Homeland Security in Your Community.* http://www.iafc.org/downloads/chckweb.pdf (accessed July 24, 2005), 2001.

Kotter, John P. *Leading Change.* Boston, MA: Harvard Business School Press, 1996.

Onieal, Denis. *Professional Status: The Future of Fire Service Training and Education.* http://www.usfa.fema.gov/downloads/pdf/nfa/higher-ed/ProfStatusArticle.pdf (accessed August 25, 2005), 2004.

Sturtevant, Thomas B. *A Study of Undergraduate Fire Service Degree Programs in the United States.* http://www.bookpump.com/dps/pdf-b/112130Xb.pdf (accessed July 24, 2005), 2001.

Bibliography

Barr, Robert C., & John M. Eversole, Eds. *The Fire Chief's Handbook*, 6th ed. Tulsa, OK: PennWell Corporation, 2003.

Bok, Sissela. *Lying—Moral Choice in Public and Private Life*. New York: Vintage Books, 1999.

Brannigan, Vincent. The real hazard is driving the fire truck. *Fire Chief Magazine*, July 1997.

Bruce, Neil. *Public Finance and the American Economy*. Reading, MA: Addison-Wesley, 1998.

Bruegman, Randy R. *The Chief Officer: A Symbol Is a Promise*. Upper Saddle River, NJ: Brady, Prentice Hall, 2005.

Bruegman, Randy R. *Exceeding Customer Expectations*. Upper Saddle River, NJ: Brady, Prentice Hall, 2003.

Carter, Harry R. Why are we dying? *Firehouse*, August 2001, p. 21.

Coleman, Ronny J. *Going for Gold: Pursuing and Assuming the Job of Fire Chief*. Albany, NY: Delmar, 1999.

Davis, Larry. Dollars & sense: The inadequacies of rural fire protection. *FireRescue Magazine*, November 2002, pp. 77–84.

Drennan, Vina. Forty more are alive today. *The Voice*, November-December 1998, p. 2.

Edwards, Steven. *Fire Service Personnel Management*. Upper Saddle River, NJ: Pearson Prentice Hall, 2005.

Emergency Management Institute, FEMA. *National Incident Management System, IS-700-Independent Study Course*. Emmitsburg, MD: USFA, 2005.

Federal Emergency Management Agency. *America at Risk, America Burning Recommissioned*. http://www.usfa.fema.gov/applications/publications (accessed September 22, 2005), 2002.

Federal Emergency Management Agency. *Responding to Incidents of National Consequence: Recommendations for America's Fire and Emergency Services Based on the Events of September 11, 2001, and Other Similar Incidents*. http://www.usfa.fema.gov/downloads/pdf/publications/fa-282.pdf (accessed September 21, 2005), 2004.

Firefighter's Handbook: Essentials of Firefighting and Emergency Response. Albany, NY: Delmar, 2000.

Fire Officer: Principles and Practice. Sudbury, MA: Jones and Bartlett, 2005.

Fisher, Roger, & William Ury. *Getting to Yes: Negotiating Agreement without Giving In*. London: Arrow Business Books, 1997.

Florida Fire Chiefs Association. *Code of Ethics*. http://www.ffca.org/ffca_administration_ethics.htm (accessed August 3, 2005), 2003.

Garvey, Gerald. *Public Administration: The Profession and the Practice*. New York: St. Martin's Press, 1996.

General Accounting Office (GAO). *Tax Gap: Many Actions Taken, but a Cohesive Compliance Strategy Needed*. GAO GGD-94-123. http://www.gao.gov (accessed August 6, 2005), 1994.

Goldfeder, Billy. Local "squeaking": The responsibility of firefighter training. *Fire Engineering*. http://fe.pennnet.com (accessed January 21, 2004), 2004.

Goodson, Carl, Barbara Adams, & Marsha Sneed. *Fire Service Ventilation*, 7th ed. Stillwater, OK: Fire Protection Publications, 1994.

Goodson, Carl, & Marsha Sneed, Eds. *Fire Department Company Officer*, 3rd ed. Stillwater, OK: Fire Protection Publications, Oklahoma State University, 1998.

Home Fire Sprinkler Coalition. *Fire Sprinkler Facts.* http://www.homefiresprinkler.org/hfsc.html (accessed May 8, 2003), 2003.

Insurance Service Office. *Fire Suppression Rating Schedule.* Jersey City, NJ: Author, 2003.

Insurance Service Office. *ISO's PPC Program: Helping to Build Effective Fire-Protection Services.* Jersey City, NJ: ISO Properties, 2001.

International Association of Fire Chiefs. *Leading the Way: Homeland Security in Your Community.* http://www. iafc.org/downloads/chckweb.pdf (accessed July 24, 2005), 2001.

International Association of Fire Chiefs & International Association of Fire Fighters. *IAFF/IAFC 2 In/2 Out Questions and Answers.* Fairfax, VA: IAFC, 1998.

International Association of Fire Chiefs & International Association of Fire Fighters. *NFPA 1710: A Decision Guide.* Fairfax, VA: IAFC, 2002.

International Association of Fire Chiefs & National Fire Protection Association. *Fundamentals of Fire Fighter Skills.* Sudbury, MA: Jones and Bartlett, 2004.

International Association of Fire Chiefs & National Fire Protection Association. *Preincident Planning Online.* Sudbury, MA: Jones and Bartlett, 2005.

International Association of Fire Fighters. *A Guide to the Recognition and Prevention of Occupational Heart Disease for the Fire and Emergency Medical Services.* Washington, DC: Author, 2001.

International Fire Service Training Association. *Chief Officer,* 2nd ed. Stillwater, OK: Oklahoma State University, Fire Protection Publications, 2004.

International Fire Service Training Association. *Essentials of Fire Fighting.* Stillwater, OK: Oklahoma State University, Fire Protection Publications, 1998.

Janing, Judy, & Gordon M. Sachs. *Achieving Excellence in the Fire Service.* Upper Saddle River, NJ: Brady, Prentice Hall, 2003.

Karter, Jr., M. J. *U.S. Fire Department Profile Through 2003.* Quincy, MA: NFPA, January 2005.

Karter, Jr., Michael J., & Joseph L. Molis. *U.S. Firefighter Injuries, November 2004.* Quincy, MA: NFPA, 2004.

Klaene, Bernard J., & Russell E. Sanders. *Structural Fire Fighting.* Quincy, MA: NFPA, 2000.

Kotter, John P. *Leading Change.* Boston, MA: Harvard Business School Press, 1996.

Lindblom, C. E., & E. J. Woodhouse. *The Policy-Making Process,* 3rd ed. Upper Saddle River, NJ: Prentice Hall, 1993.

Manning, Bill. Editor's opinion—Blood on their hands. *Fire Engineering,* November 2002.

Maryland Fire & Rescue Institute. *Fire Department Company Officer Curriculum.* College Park, MD: Maryland Fire & Rescue Institute, 2004.

National Commission on Fire Prevention and Control. *America Burning: The Report of the National Commission on Fire Prevention and Control.* http://www.usfa.fema.gov/usfapubs/pubs_main.cfm (accessed September 19, 2005), 1974.

National Fire Academy. *America Burning Revisited.* Washington, DC: U.S. Fire Administration, 1987.

National Fire Safety and Research Office. *Introductory Summary: Fire Prevention and Control Master Planning.* Washington, DC: U.S. Department of Commerce, National Fire Safety and Research Office.

Neustadt, Richard E. *Presidential Power: The Politics of Leadership.* New York: Wiley, 1990.

NFPA. *Codes, Standards, and Recommended Practices.* Quincy, MA: National Fire Protection Association. (*See www.nfpa.org for latest editions.*)

NFPA 1, *Uniform Fire Code*

NFPA 299, *Standard for Protection of Life and Property from Wildfire*

NFPA 471, *Recommended Practice for Responding to Hazardous Materials Incidents*

NFPA 472, *Standard for Professional Competence of Responders to Hazardous Materials Incidents*

NFPA 1001, *Standard for Fire Fighter Professional Qualifications*

NFPA 1002, *Standard for Apparatus Driver/Operator Professional Qualifications*

NFPA 1003, *Standard for Airport Fire Fighter Professional Qualifications*

NFPA 1021, *Standard for Fire Officer Professional Qualifications*

NFPA 1031, *Standard for Professional Qualifications for Fire Inspector and Plan Examiner*

NFPA 1033, *Standard for Professional Qualifications for Fire Investigator*

NFPA 1035, *Standard for Professional Qualifications for Public Fire and Life Safety Educator*

NFPA 1041, *Standard for Fire Service Instructor Professional Qualifications*

NFPA 1051, *Standard for Wildland Fire Fighter Professional Qualifications*

NFPA 1061, *Standard for Professional Qualifications for Public Safety Telecommunicator*

NFPA 1201, *Standard for Developing Fire Protection Services for the Public*

NFPA 1221, *Standard for the Installation, Maintenance, and Use of Emergency Services Communication Systems*

NFPA 1403, *Standard on Live Fire Training Evolutions*

NFPA 1404, *Standard for Fire Service Respiratory Protection Training*

NFPA 1410, *Standard on Training for Initial Emergency Scene Operations*

NFPA 1451, *Standard for a Fire Service Vehicle Operations Training Program*

NFPA 1500, *Standard on Fire Department Occupational Safety and Health Program*

NFPA 1521, *Standard for Fire Department Safety Officer*

NFPA 1561, *Standard on Emergency Services Incident Management System*

NFPA 1620, *Recommended Practice for Pre-Incident Planning*

NFPA 1710, *Standard for the Organization and Deployment of Fire Suppression Operations, Emergency Medical Operations, and Special Operations to the Public by Career Fire Departments*

NFPA. *Firefighter Fatalities Due to Sudden Cardiac Death, 1995–2004*. Quincy, MA: NFPA Fire Analysis and Research, 2005.

NFPA. *A Needs Assessment of the U.S. Fire Service. A Cooperative Study*. Authorized by U.S. Public Law 106-398: FEMA & NFPA. http://www.nfpa.org/assets/files/PDF/needsassessment.pdf (accessed July 21, 2005), 2002.

NFPA. *NFPA Fire Protection Handbook,* 19th ed. Quincy, MA: NFPA, 2003.

NFPA. *State-by-State Findings of Fire Department Needs & Response Capability*. Quincy, MA: NFPA, 2004.

NFPA. *U.S. Firefighter Fatalities in the United States*. Quincy, MA: NFPA, June 2004.

O'Hagan, John T. Staffing levels, part 5. *Fire Command*, March 1985, pp. 18–21.

Onieal, Denis. *Professional Status: The Future of Fire Service Training and Education*. http://www.usfa.fema.gov/downloads/pdf/nfa/higher-ed/ProfStatusArticle.pdf (accessed August 25, 2005), 2004.

Page, James. A taboo topic: Fire service professionals ignore a big problem. *Fire-Rescue Magazine*, October 2002, p. 10.

Page, Jim. A master at his art. *Fire-Rescue Magazine*, May 1998, p. 27.

Perkins, Ken, & John Benoit. *The Future of Volunteer Fire and Rescue Services: Taming the Dragons of Change*. Stillwater, OK: Oklahoma State University, Fire Protection Publications, 1996.

Phoenix Fire Department. *1998 Phoenix Fire Department National Survey on Fire Department Operations*. Unpublished, 1998.

Powell, Colin. *Quotations from General Colin Powell: A Leadership Primer—18 Lessons from a Very Successful American Leader*. http://www.usna.edu/JBHO/sea_stories/quotations_from_general_colin_powell.htm (accessed May 17, 2003), 1996.

Resources for Fire Department Occupational Safety and Health. Quincy, MA: NFPA, 2003.

Riley, Bryan, Eric V. Schlecht, & John Berthoud. *Hidden Taxes: How Much Do You Really Pay?* Institute for Policy Innovation (IPI). http://www.ipi.org (accessed August 6, 2005), August 29, 2001.

Schick, Allen. *The Stages of Budget Reform in Government Budgeting: Theory, Process, Politics: The Road to PPB.* Paris: OECD, 1995.

Smoke, Clinton H. *Company Officer,* 2nd ed. Clifton Park, NY: Thomson Delmar Learning, 2005.

Snook, Jack W., & Jeffrey D. Johnson. *Cooperative Service Through Consolidations, Mergers and Contracts—Making the Pieces Fit.* Stillwater, OK: Oklahoma State University, Fire Protection Publications, 1997.

Snook, Jack W., Jeffrey D. Johnson, & Dan C. Olsen, with John Buckman. *Recruiting, Training, and Maintaining Volunteer Firefighters.* Stillwater, OK: Oklahoma State University, Fire Protection Publications, 1997.

State Farm. *Frequently Asked Questions About Subzone Rating.* Unpublished paper. Bloomington, IL: Author, 2001.

U.S. Department of Homeland Security (DHS). *National Incident Management System.* http://www.dhs.gov/interweb/assetlibrary/NIMS-90-web.pdf (accessed August 13, 2005), 2004.

U.S. Fire Administration, FEMA. *Fire in the United States 1992–2001,* 13th ed. http://www.usfa.fema.gov/statistics/reports/pubs/fius13th.shtm#accessible (accessed August 21, 2005), 2004.

U.S. Fire Administration, FEMA. *Guide to Developing Effective Standard Operating Procedures for Fire and EMS Departments.* Emmitsburg, MD: FEMA, 2000.

U.S. Fire Administration, FEMA. *Guide to Funding Alternatives for Fire & EMS Departments.* Emmitsburg, MD: FEMA, 2000.

U.S. Fire Administration, FEMA. *Recruitment and Retention in the Volunteer Fire Service.* Emmitsburg, MD: FEMA, 1998.

Wilson, Paul F., Larry D. Dell, & Gaylord F. Anderson. *Root Cause Analysis: A Tool for Total Quality Management.* Milwaukee, WI: ASQC Quality Press, 1993.

Zagaroli, Lisa. Ambulance trips put lives at risk. *Detroit News,* January 26, 2003, Front page.

Index

human resources management. *See also* staffing
empowering employees, 174, 182
insubordination, 84–86
job analysis and classification, 87–89
motivation, 93

I

IBC (International Building Code), 134–135
ICC (International Code Council), 134
identifying problems, 36–39, 43
impact fees, 58
implementation
of change, 45–46
public policy, 167–168
IMS (incident management system), 6–7, 119, 185–186
in-service training, 107. *See also* education and training
incident management system, 6–7, 119, 185–186
income and funding, 49–52, 57–58. *See also* finances and economics
alternative sources, 64–65
auditing and accountability, 68–69
estimating, 58
from fees. *See* fees
protecting the budget, 69–70
from taxes. *See* taxes
voting and public choice, 55–56
income taxes, 50, 60
increasing productivity, 159
inequity, tax, 59–60, 62
influence. *See* consensus building; power
informal organizations, 26–28
information gathering, 29
feedback from staff, 29–30, 43
MBWA (managing by walking around), 25–26
initial fire attack performance, 114
injuries on the job (IOJ), 80, 118, 125. *See also* health and safety
inspections, 51, 103
institutionalizing change, 45–46
insubordination, 84–86
Insurance Service Office grading schedule, 2–3, 170–173
interactive multimedia training, 108
interest groups, 44, 131
internal audits, 68
International Building Code (IBC), 134–135

International Code Council (ICC), 134
Internet-based programs, 109
investing, 52
IOJ (injuries on the job), 80, 118, 125. *See also* health and safety
ISO Grading Schedule for Municipal Fire Protection, 2–3, 170–173

J

job analysis and classification, 87–90
joint purchase, 65–66
judgment, 41
judicial system, 142–144
jumping over barriers, 43–44
justification
budget decisions, 54
ethical, 146–147
regulatory legislation, 97, 128–129

K

Kimel et al. v. Florida Board of Regents, 79
Klinger, Keith E., 71
knowledge development, 108–109

L

labor relations, 92. *See also* human resources management
labor unions, 91–93, 135–136
law. *See* legal issues
lawsuit costs, 78
lawyers, 143
leadership, 15–16, 20, 46
in change, 33–46
fire chiefs. *See* chiefs
insubordination, 84–86
leading by example, 42, 46
networking. *See* networking
troublesome supervisors, 44
vision. *See* vision
lease-purchase financing, 52
legal issues
FLSA (Fair Labor Standards Act), 83
human resources, 77–80, 86–87
justification for government involvement, 97, 128–129
process and court issues, 142–144

Credits

Chapter 1
1-1 Courtesy of Aurora Regional Fire Museum; 1-2 Courtesy of Dennis Chesters, Laboratory for Atmospheres, NASA Goddard Space Flight Center; 1-3 © Jones and Bartlett Publishers. Courtesy of MIEMSS.

Chapter 2
2-1 Courtesy of Lance Cpl. Brian Kester/U.S. Marines; 2-2 Courtesy of L. Charles Smeby, Jr.

Chapter 3
3-1 Courtesy of Ocean City-Wright Fire Department (Chief Randy Brown); 3-2 Courtesy of Paul O. Davis/On Target Communications

Chapter 4
4-2 Courtesy of Michael Rieger/FEMA

Chapter 5
5-1 Courtesy of L. Charles Smeby, Jr.; 5-2 © Jason Maehl/ShutterStock, Inc.

Chapter 6
6-1 © Joy Fera/ShutterStock, Inc.; 6-2 Courtesy of NOAA; 6-3 Courtesy of L. Charles Smeby, Jr.

Chapter 7
7-1 © Glen E. Ellman; 7-2 Courtesy of L. Charles Smeby, Jr.; 7-3 Reprinted Courtesy of Massachusetts Department of Fire Services; 7-4 Courtesy of L. Charles Smeby, Jr.

Chapter 8
8-1 © Glen E. Ellman; 8-2 © Peter Willott, St. Augustine Record/AP Photo

Chapter 9
9-1 Courtesy of National Right to Work Legal Defense Foundation, Inc. Used with permission.; 9-2 © Index Stock Images, Inc./Alamy Images

Chapter 10
10-1 © www.cartoonstock.com

Chapter 11
11-1 Courtesy of City of Baltimore, Office of the Mayor

Chapter 12
12-1 © Jones and Bartlett Publishers. Courtesy of MIEMSS.

Unless otherwise indicated, all figures are under copyright of Jones and Bartlett Publishers, Inc.